FOUNDATIONS

for Personal & Professional
Project Management

an effective step-by-step guide to
take ownership of your
project management processes

Copyright © 2010, 2012, 2019 by Dr. Dennis E. Odiaka, PMP, SMC.
Foundations for personal and professional project management

All rights reserved. No part of this publication may be reproduced or transmitted in any form or by any means—electronic or mechanical, including photocopying, recording, or by any information storage and retrieval system without permission in writing from the publisher. The publisher authorizes no abridgment or changes to the text.

Contact Author: mail@countinstitute.com

ISBN: 978-1-79473-448-7

Printed in the United States of America

Foundations
for Personal and Professional Project Management

BOOK 1

Dr. Dennis Odiaka, PMP, SMC.

Dedication

I heard that the beauty of the original is in the originality of beauty.
I also heard that wisdom is originally God's.
I can tell that the beauty of wisdom is in its original use.
I can also tell that God's use of project management methods during creation is originally wise since creation—the deliverable—of His creative work shows off the beauty of wisdom.

Therefore, I dedicate this book to those who seek wisdom, originality, and concomitant beauty in the knowledge of project management.

Acknowledgment

It may be difficult to comprehend—and for some, accept how a dark and disarranged problem of our cosmos—the earth, was solved in seven days. Actually, it was in six days. However, that six-day unique and temporary endeavor of God brought permanence of light and arrangement. Such that we can momentarily pause and admire the relative simplicity of existence, even in the face of so many complexities behind the scene.

As a result, I infer that the degree of simplicity of a thing is directly proportional to the degree of complexity of that thing. For me, project management is a simple endeavor if you know what you are doing. Nevertheless, it is a cosmos of relative complexities.

I began the project of writing this book by thinking of how to communicate, simply, the manifold wisdom of project management. Therefore, I thank God and glorify Him for being the woodcutter and the first inspiration. I am also gratefully appreciative of my wife and family for their continued support toward my quest for finding better—if not simpler—ways of doing things. Sandi Huddleston-Edwards—project manager, author, and English teacher—edited this work and she has been a blessing. Her words of mean plenty and I thank her for everything.

I have had to bounce ideas back and forth with some people in order to stay the course: Matt Njoku—Assistant Professor of Finance & Economics and Director of Entrepreneurship Education at Montreat

College; Dr. B. Bayode—Associate Program Dean at Strayer University; and the countless number of people who may not realize they contributed significantly towards the birthing of the book. I owe them many thanks. I will always remember Dr. Ahmad of Walden University for the many words of professional writing counsel he has given.

Importantly, I must acknowledge those who seek project management knowledge as represented by my Information Technology and Project Management students at the University of Phoenix, Capella University, and Devry University. I also acknowledge my future project management students, including the ones I met at Montreat College, where I was director of the master of science in management and leadership graduate program in the School of Professional and Adult Studies. All my students opened a window of opportunity for me to see areas where people and professionals lacked or needed for knowledge to be more simplified. I hope I have been able to represent simply the complex body of knowledge that is project management in this book.

Preface

    We learn from the Chinese that the best way to eat an elephant is to do it one bite at a time. As long as you are sure, the elephant is not alive. This book has adopted this approach in presenting the humongous body of knowledge known as project management so that acquiring project management competency should be as easy and enjoyable as eating a humongous elephant, one bite at a time. The prophets of old call this strategy a line-upon-line, precept-upon-precept approach.

    The idea of FOUNDATIONS is to establish you, firmly, on the BASIC and CORE principles that guide the practice of project management. This is so that you can competently take OWNERSHIP of the knowledge of every process and area of project management. It is a fact that the project management discipline has become one of the most sought after professional practices today. This is because it has evolved to become that effective and well-organized professional practice that our organizations need to effect CHANGE.

    Change is a phenomenon we cannot ignore. It drives everything we do. A preacher once said that people are created for movement. Scientists also say the same of particles being in constant movement. Time is never constant, as our world is perpetually changing from one season to another. Our environment does not accommodate immovability for long. We must change to survive, as we must transform to prosper. For managers and professionals, the cost of change may be a few thousand dollars. But what is the benefit of receiving training in a skill set that facilitates change? Priceless.

Project management is a change agent. Those who possess the skill and capability for managing projects professionally own arsenals of competitive advantage. These include (a) increased job opportunities that require formal knowledge in project management, (b) advancement in the profession of project management and expansion of knowledge base, (c) significant improvement of project management capabilities, (d) acquisition of a structured discipline, which has proven successful for other project managers, and (e) enhancement of earning-power and others.

This book is a concise overview of the project management body of knowledge. It previews, presents, and decomposes project management as a body of knowledge into manageable learning packages. These packages are relevant to provide education, development, and training for the acquisition of competence. The structure of presentation follows an examination of the beginning, evolution, and principles to a synthesis of application areas of project management.

Table of Contents

(BOOK 1)

Division 1
Project Management Trenches and Foundations   1

Division 2
Project Management Cornerstones   51

Division 3
Project Management Stonecutters and Implements   95

Not included in this cover.

(BOOK 2)

Division 4
Project Management Stone Walls   131

Division 5
Project Management Capstone

Table of Contents

Division 1—Project Management Trenches and Foundations 1
Chaapter 1: Managing Projects 3
   In The Beginning 4
   Project Management Today 5
   Project Management Challenges and Resolution 7
   What is Project Management? 10
      Projects 11
      Key Characteristic and Definition of a Project 11
      What is Not a Project? 14
   Test Your PM Knowledge 17

Chapter 2: Defining Project Management 19
   Management 20
   Project Management 21
      Project Management Objectives and Benefits 21
      Project Management Functions 23
   The Systemic World of Project Management 25
      The Systems Theory 25
      Parts and Entities of Project Management Systems 26
      Project Teleology and Equifinality 28
   Test Your PM Knowledge 31

Chapter 3: Project Management CAKES—Contexts, Application areas, Knowledge areas, Environment, & Skills 33
   Project Management is not a Cakewalk, or is it? 34
   Project Management Broader Contexts 34
      Subprojects 35

  Program Management 36
  Portfolio Management 37
  Project Management Office (PMO) 37
 Project Management Socio-Economic Environment 40
 Project Management Body of Knowledge..42
 Interpersonal Skills and General Management Knowledge 43
 Test Your PM Knowledge 49

Division 2—Project Management Cornerstones 51
Chapter 4: Project Life Cycle, Phases, and Process Groups..................................53
 Project Life Cycle 54
  Industry Preferred Project Life Cycles 55
 Phases and Stages of a Project 56
 Project Management Phases and Common Characteristics 58
 Process Groups–Initiating, Planning, Executing, Monitoring and Controlling, Closing
64
  What is a Process?..66
  The Five Process Groups 68
  Initiating 68
  Planning..69
  Executing 70
  Monitoring and Controlling 71
  Closing 71
  Interaction of Process Groups 71
 Test Your PM Knowledge 73

Chapter 5: Managing Small and Simple, Large and Complex Projects 75
 Type of Project Distinction 76
 Small and Simple Projects 76
  Managing Small and Simple Projects..78
 Managing a Project Using Project Phases and Process Groups 80
  Construct a House Model and Build the House.......................................80
  Project Failure and Lessons learned 86
 Large and Complex Projects 88
 Test Your PM Knowledge 93

Division 3—Project Management Stonecutters and Implements 95
Chapter 6: Project Tools and Techniques...97
 Project Management Tools and Techniques 98
 The Gantt Chart 99
  Uses, Benefits, and Limitations of Gantt Charts 101

Flowcharts	101
Uses, Benefits, and Limitations of Flowcharts	103
WBS, Decomposition	103
Uses, Benefits, and Limitations of WBS	106
Test Your PM Knowledge	107
Chapter 7: Project Management Technology	109
Project Logbook	110
Project Management Software	111
Types and Categories of Project Management Software	112
Test Your PM Knowledge	115
Chapter 8: Project Management Influences	117
Project Organization	118
Projects Triple Constraints	123
Project Processes and Knowledge Activities Interaction	125
Test Your PM Knowledge	127
Bibliography	129
Glossary	131

Book 2:
Division 4—Project Management Stone Walls
   Chapter 9: Integration Management
   Chapter 10: Scope Management
   Chapter 11: Time and Schedule Management
   Chapter 12: Cost and Budget Management
   Chapter 13: Quality Management
   Chapter 14: Human Resources Management
   Chapter 15: Communications Management
   Chapter 16: Risk and Uncertainty Management
   Chapter 17: Procurement Management
   Chapter 18: Project Management Professional Responsibility

Division 5—Project Management Capstone:
   (Skill Development and Practice)
   Chapter 19: Project Management Documentation
   Chapter 20: Workshop and Practice

Division 1

Project Management Trenches and Foundations

Chapter 1

Managing Projects

- *In The Beginning*
- *Project Management Today*
- *Project Management Challenges and Resolution*
- *Defining Project Management*

In The Beginning

Project management is an ancient yet very contemporary profession. It is as ancient as civilization, having been used for massive projects like the Great Pyramid of Giza, the Coliseum of Rome, the Mayan temples, the Taj Mahal, the Great Wall of China, and others. For example, the Great Pyramid of Giza constructed in 2570 BC—4578 years ago, ranked as the tallest human-made structure on Earth for over 38 centuries—from 2570 BC to 1300 AD and stood at 146 meters compared to Korea World Trade Center in Seoul constructed in 1988 and stands at 227 meters.

These colossal ancient projects required massive amounts of resources. The edifice took over seven million person weeks, more than twenty years, and over two million blocks of stone. By every modern-day standard, a great deal of exceptional planning, management of labor and logistics, deployment of materials, and money went into constructing these ancient projects. Granted ancient projects were successful. Management techniques used for executing these projects are archaic. This is in the sense that their language, communication, and understanding of projects among others lacked the state-of-the-art technology available to us today.

> As of April 7 2008, The **Burj Khalia** (formerly **Burj Dubai**) in Dubai, United Arab Emirates currently stood as the world's tallest structure at 629m (2,064 ft). Since 2010 it rose to over 829m (2,717 ft).

Notwithstanding, project management in ancient times was premised on the desires, resources, and self-serving needs of Kings, Princes, and Royals. As a result, ancient project management lacked the care, concern, and limitation today's professionals and practitioners have about costs, completion time, and human lives.

Project Management Today

Societal and environmental changes immensely contributed to the evolution of project management. More importantly, the need for change that has driven how project management is practiced today. This is because projects are about change, which in various forms help us to alter, modify, and transform our environment.

As isolated societies evolved from primarily creating tools for survival to creating tools for increasing productivity in complex and integrated work environments, managerial problems led to the development of what we know today as project management. Although some early projects were definitive on scope, they were not definitive on time and cost considerations. The advent of the scientific and industrial age influenced the considerations given to project scope, time, and cost. These three components (or constraints) of project management helped factories create wealth and marketplace for improved standards of living, expanded trade, and competition.

> **Project Scope, Time**, and **Cost** are components of project management that respectively define (1) The basis and details for the expectations of a specific project deliverable or end result; (2) The amount of time required to complete a project even though it is not considered a resource or cost; and (3) The resource, labor, and material costs required to complete a project.

The development of quality management in the 1950s saw another relatively subjective component of project management emerge. The quality management strategy emerged as a significant response to sustain organizational effort through a customer focus orientation. This orientation led to the integration of quality in project management, as it became a basis for competition, survival, and profitability of organizations.

Considering all four parameters, the form and function of modern-day project management enable us to act in specific creative and vigorous ways that continuously improve societal and organizational goals. As societal evolution continued through manufacturing to knowledge-based organizations in the information age, more management parameters were included as support systems for the three components already mentioned. These parameters, which include communication, integration, risk, human resources, and procurement management, were consequent on the influence of advanced technology, such as communication and information technologies.

> **Project Management Support Systems** for the triple constraints components include communication, integration, risk, human resources, and procurement management. These in addition to project cost, time, and scope constitute the Project Management Knowledge Areas.

It requires tremendous personal and professional effort and ability to develop an admirable skill set for managing projects successfully. This is because managers particularly are pressured to keep costs down, work in less time, not change scope, and not compromise quality. Nonetheless, the project management skillset is one of the fastest-growing professional disciplines in North America and the world at large. Therefore, this makes it a highly demanded commodity. Project management systematically decomposes the chaos of an overwhelming project workload into manageable elements of scope, time, and cost, quality, human resources, communication, risk, procurement, and integration. Within the last decade, project management has grown extraordinarily as a discipline that has aided our societies to continue to master the complexity of its environment.

Project Management Challenges and Resolution

Like you, managers and professionals are challenged continuously by the need to manage projects effectively for success, much so by the pressure to reduce *costs* and *time*, while not compromising on *scope* and quality. This is not an impossible task, but a difficult one. Managers, therefore, seek ways to acquire relevant project management training, capability, and competency.

Figure 1.1 –Project Management Three Constraints Triangle

The reason for such a challenge is that most organizations today engage in continuous improvement, change, and development. They want to deliver full baseline **scope** at minimal **cost** within the shortest possible **time**. The challenges result from a rapidly changing, complexly, evolving global and highly competitive environment. They must survive and grow in the face of domestic and international competition.

> Early and Currently Used *Project Management Tools & Techniques*:
> 1. Gantt Chart
> 2. Flow Chart
> 3. WBS
> 4. CPM
> 5. ADM
> 6. PERT

As a function of sudden or gradual transformation, operational and organizational change involves the whole or parts of any organization. In order to achieve effective change, project management as a body of knowledge provides knowledge, processes, tools, and techniques that are scalable and adaptable for use in a variety of operations, organizations, or industries. Such tools and techniques include the following:

1. **Gantt Charts**—Industrial engineering techniques developed by Henry Gantt in 1910 to show and monitor planned project activities and relationships.

2. **Flow Charts**—A schematic representation of an algorithm or a process introduced to ASME (American Society of Mechanical Engineers) members in 1921. It is also one of the basic tools of quality management.

3. **WBS**, the Work Breakdown Structure—A technique for defining and organizing the total scope of a project using sets of planned outcomes in a hierarchical tree structure.

4. **CPM**, the Critical Path Method—A network diagram for identifying the amount of schedule flexibility and minimum project duration.

DuPont and the United States Navy developed the CPM after World War II. It is used to calculate the earliest possible completion date for a project based on the sequencing and duration of activities.

5. **ADM**, the Arrow Diagramming Method—A variant of the CPM uses arrows to represent project activities and circular nodes to represent events at the start and end of project activities.

6. **PERT**, the Program Evaluation and Review Technique—A network modeling application originally designed in 1958 by the U.S Navy for planning and controlling its Polaris nuclear submarine project.

Project Management (PM) Development Timelines:

1950: Marked the beginning of modern day PM era and the creation of PM tools & methods.
1960: PM tools & methods were integrated with the computer.
1967: IPMA—the International Project Management Association was formed in Europe.
1969: PMI—the Project Management Institute was formed in the U.S.
1970: PM concepts expand to the construction, information and other industries.
1980: PM use expands further, and it became the fastest growing discipline.
1981: PMI Board of Directors authorized the development of PMBOK Guide.
1986: PMI certified the first Project Manager.
1987: PM Professional Associations began forming around the world.
1997: PM growth led to the addition of 'Integration' to the PM Knowledge Areas.
2008: There are over 100,000 certified Project Management Professionals.

In general, project management will help you learn and employ a structured and disciplined approach to managing projects. This is so that you are able to balance specific projects need with the project management methodology. Consequently, you and your organization can strive to succeed and establish competency at project management. However, the appropriate strategy for organizations is to implement and practice standardized procedures that regularize project management at every level of the organization. However, there is more about project management than the sketchy premise we have painted of it. The question is, what exactly is project management in modern parlance?

What is Project Management?

The right approach to answer the question of what is project management is to consider its complex and systemic nature. What this means is that we will decompose the term project management. Decompose is a popular project management language, so it is appropriate that we use it to unhinge the compound term known as project management for easy understanding.

Projects and *Management* combine to form the term project management. So what is a project? We have already discussed that projects are about and facilitate change. That is because they help us alter, modify, or transform the present state of things to where we want them.

> **A project** is a one-time temporary and unique endeavor that has definite goals and outcomes, and utilizes budgets and schedules, human resources and standards in the face of uncertainties to create change through deliverables within specific periods.

Projects

Projects vary in composition and objectives. They are what we do all the time, every time. Reading this book might mean doing a project. The cup of tea you made yourself during that cold morning some time ago was a project. On the other hand, that dream house model you or someone you know built is a project. It began by having a desire—dream, vision, or need—for a cup of tea. The fulfillment of that desire led to the process of gathering the right kind of resources to convert the vision of a nice hot cup of tea to reality. The vision became real when that steaming cup of tea in your hands found its way to your lips for a well-deserved warm sip.

The simplicity of the concept of a project might sound strange, but the basic principle is in what we do all the time. Therefore, the answer to what a project means lies in our awareness of the power of a project to enable us to use resources efficiently and effectively to create things or accomplish tasks. Beyond the simplicity of its concept, we use projects to plan, direct, and execute ways to use organizational resources more effectively. A project, therefore, is, without a doubt, the answer to needs. It supposedly acts as a means through which you or your organization can focus resources and abilities towards accomplishing a task or desired outcome. A project helps us achieve our goals.

Key Characteristics and Definition of a Project

To this end, we can characterize a project as a small or large, simple or complex undertaking by anyone and at any level in an organization, which may involve one or more people or take one or more days to complete. However, the fundamental characteristic of a project is that it is a one-time endeavor, just like making a cup of tea or building a house.

According to *Webster's 2005 New World Dictionary & Thesaurus*, a project is a proposal of something to be done; a plan; a scheme; an organized undertaking; a special unity of work, research; or a public undertaking, as in conservation or construction. While this definition is not adequate to convey the totality of the meaning of a project, let us consider the definition provided in Project Management Institute's *A guide to the project management body of knowledge: PMBOK® guide*– 3rd Ed.

> **Operations** are used for routine, repetitive, and continuous work in organizations where deadlines are not crucial and operations procedures are standard.
>
> **Projects** are used for unique one-time undertakings and functions in organizations, where deadlines are crucial.

PMBOK® Guide–3rd ed. defines a project as a temporary endeavor undertaken to create a unique product, service, or result. Regardless of the size, cost, and outcome of a project, the operative words here are creating, temporary, unique, and results. From these and the foregoing, we can extract characteristic fundamentals about a project, which leads us to conclude that

1. **Projects are About Change**—whether small or large, temporary or permanent, trivial or significant, all projects are about change as they create new things to displace the old.

2. **Projects are One-time Endeavors**—they are one-off events that come and go leaving behind the product(s) of their endeavor.

3. **Projects are Temporary**—they have a definite beginning and a definite end, undertaken to achieve a definite goal and objective and ends when the project objectives are accomplished.

4. **Projects are Unique**—small or large, simple or complex, similar or different in appearance; every project at its core is unique and never identical to another project.

5. **Projects are not forever**—projects are about creating things within specific times, deadlines, targets, or completing dates. They have limited and defined periods: a definite beginning and a definite end.

6. **Projects have defined Goals, Objectives, and Outcomes**—all projects have well defined measurable and achievable deliverables relative to their goals and objectives.

7. **Projects have Budgets and Schedules**—cost and schedule constraints affect the execution of all projects.

8. **Projects have Risks**—known and unknown risks, as well as uncertainties, face all projects, which may jeopardize the successful completion of any project.

9. **Projects Involve People**—from one individual to a nation of people, projects are executed by people regardless that they are challenged with limited availability of human resources.

10. **Projects use Standards**—at the beginning of any project, standards are defined to keep projects conforming to quality.

11. **Projects Grow in Steps and Details**—as more detailed and specific information becomes available, they are used to elaborate a project by detailing and planning, progressively.

12. **Projects have a Life Cycle**—all projects follow a pattern as they progress from initiation to completion in definable phases.

> *Examples of Projects:*
>
> Projects can be used for the following and more:
> - To develop a new product or service, such as a new transportation or voting system,
> - To effect change is an organization's process or structure,
> - To construct a new building or develop a new computer software,
> - For strategic planning to address issues of market demand, customer request, or technological advance.

What is Not a Project?

It may certainly seem easy at this point to describe what a project is not. It can also be confusing if a project is expected to fulfill a variety of needs. To put it in context, you work in an organization that develops new products and services. Granted that we have established that projects create products, we certainly concur that projects run the gamut of product development projects of your organization from the very simple to the very complex. There will certainly be a fine line between your daily work functions and your daily project work, taking into account how projectized your organization maybe. The question here is how you would differentiate between what is and what is not a project?

All work performed in and by organizations is done to achieve a set of objectives. Such works are categorized as either operations or projects, both with clear distinctions yet similar characteristics. Projects may vary significantly depending on organizational objectives. However, they are classified along the lines of small or very small, large, or very large, simple or complex. They can also be undertaken by one or more people, executed at any level of the organization in any department or workgroup for short or lengthy periods, involve joint ventures or partnerships, and are in any or multiple locations.

Table 1.1
Difference between Projects and Operations:

Project Work	Operations Work
Develops new products or service	Help to sustain an existing business
Is more unique	Is less unique
Is varied	Is about routine work
Is used for effecting change	In not involved in introducing change
Terminates after the goal is achieved	Continues with new objectives
Is used in project management	Is used in operations or production management
Is planned and tracked	Uses standard operational procedures
Is orientation is for the future	Is oriented for the present
Seeks standardization and efficiency of work	Seeks the effectiveness and success of operations
Creates new things	Maintains existing things
Organizes activities outside of operational limits	Organizes activities within its operational limits

The functions of projects and operations in organizations sometimes overlap, they also share similar characteristics, such as they are planned, executed, controlled; performed by people; and constrained by limited resources. However, they are characteristically different. Table 1.1 lists their fundamental difference.

It is the nature and uniqueness of the work that distinguishes projects from operations work—this is also known as production work in certain quarters. While the objectives of a project are to attain its pre-defined goal and terminate, the objectives of an operation are the contrary: to sustain a business continuously. At the attainment of its specific objective, a project closes, but an operation sets or adopts new sets of objectives to continue work.

17 | Managing Projects

Test Your PM Knowledge

1. True or false: Project management is an ancient trade.
    a. True
    b. False

2. Modern day practice of project management is limited by_____
    a. Cost, schedule, and time
    b. Cost, completion time, and human resources
    c. Cost, completion time and activities

3. Project scope, time, and cost are components of project management that respectively define
    a. The basis and details for the expectations of a specific project deliverable or end result
    b. The amount of time required to complete a project even though it is not considered a resource or cost
    c. The resource, labor, and material costs required to complete a project
    d. All of the above
    e. Only options (a) and (b) are correct.

4. Project management triple constraints include _____
    a. Cost, time, and scope
    b. Time, cost, and activity
    c. Scope, training, and competency
    d. Quality, time, and scope
    e. Quality, cost, and time

5. The oldest known project management tool still used today by project managers is the _____
    a. Flow chart
    b. Gantt chart
    c. PERT
    d. ADM

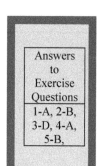

Answers to Exercise Questions
1-A, 2-B, 3-D, 4-A, 5-B,

Chapter 2

Defining Project Management
- *Management*
- *Project Management Definition*
- *The Systemic World of Project Management*

Management

We are still in the process of defining project management. Having defined projects, we will examine its other component—management. The right question to ask at this point is what is management? It is simply an activity or process of putting together and using a set of resources to accomplish a pre-determined goal or objective. In this description of management, notice that there is a direction towards a goal, an action, a process, and assembling and use of resources.

You can deduce from this description that a goal is reachable if there are a plan and some measure of control. Therefore, management can be defined as a simple or complex act of assembling human and other resources to achieve desired goals through processes of *planning, organizing, controlling, directing,* and *coordinating*. These processes are regarded as the five functions of management.

> **Management** is simply the art and act of assembling human and other resources to achieve desired goals through processes of planning, organizing, controlling, directing, and coordinating.

There is no doubt that this definition delineates the science of management. Given that a certain level of subjectivity exists anywhere humans gather, it will not be entirely apt to regard management as a science only. Therefore, management is also considered an art of getting things done through people in order to reach an intended goal. Whether as an act or art or both, one thing is sure, management is what we do—the functions we perform—as managers to reach our goals.

Project Management

If we combine our descriptions of **project** and **management**, we will have something like the following. Project management is what we do to create changes that have definite outcomes. To apply the dispositions of the art and science of management, project management is the applied art and act of managing projects. Wherefore, abstract objectives converted into concrete goals or visions become a reality.

On the position that projects are change agents and create things, project management is also seen as change management. The benefit of project management as a change agent is in the efficiency of its methods to facilitate change. However, there is more about project management than these varieties of simplistic descriptions. The Project Management Institute, in PMBOK® Guide-3rd ed., defined project management as the application of knowledge, skills, tools, and techniques to meet project requirements.

Such knowledge, skills, tools, and techniques include defining and planning the necessary work to be done, scheduling activities to complete the work, monitoring and controlling project activities, and closing or conducting activities to end the project. Project management also involves coordinating and directing the work of other people like the project manager(s) and the project team(s) involved in the project. Generally, project management has been beneficial to organizations because of the following reasons.

Project Management Objectives and Benefits

There are a variety of benefits in using project management methodology as an organizational or managerial approach to managing

> **Management by Project** is the tendency to manage operations activities in some application areas using project management methodology.

projects and some ongoing operations. Granted that not all operations can be managed as projects, some organizations define some of their operations on a project basis. Consequently, they get to define and adopt such approaches as *management by projects*. The tendency is to manage certain activities in specific application areas using the principles of project management. Nevertheless, the reasons why project management is applied to a variety of areas are because project management

- Provides the **knowledge, processes, tools, and techniques** that create discipline and ways for organizing project data, which aid the successful management of all projects regardless of their size or complexity.
- Provides a way to define **project scope**, control such scope changes, and enable effective management of project time and cost within specific application areas.
- Helps with the definition of **project roles and responsibilities**, which allows the establishment of order in a non-linear dynamic environment.
- Establishes clear and **achievable objectives**, furthers the strategic goals of an organization and manages the expectation of project stakeholders by adapting to specifications and plans.

However, it is by following and adapting to specifications, procedures, and processes that project management is able to accomplish its objectives. The specifications, processes, and procedures begin and end with

- Having a *goal,* based on customers or stakeholders' request.
- Developing a feasible *plan* to achieve the goal and following the plan.
- Selecting and using appropriate *tools and techniques* to manage useful resources so that goals are achieved on time, on budget, and within scope, while not compromising on quality and other technical standards.
- Involving and using *human resources*, that is people or persons with the right kind of knowledge and skills needed to achieve the project goal.
- Engaging and effectively managing *stakeholder's expectations* regardless of their diversity.
- Knowing and having an understanding of how to manage *uncertainties and risks* that are inherent in projects throughout all project management phases that comprise of project initiation, planning, execution, monitoring and controlling, and closing.
- Having adequate knowledge of project management *knowledge areas,* which includes project integration, scope, time, cost, quality, human resources, communication, risk, and procurement as well as professional responsibility.

Project Management Functions

We can establish from the preceding description of projects and management functions that the following functions facilitate project management. They include

- *Planning*—this requires the use of imagination or enterprise or both to develop an ambitious scheme, arrangement, contraption, or stepwise approach for doing something; in this case, a project.

- ***Scoping***—this requires the ability to define a range, boundary, or the extent of action or activity of a project in order to effectively plan activities, estimate costs, and manage expectations.

- ***Estimating***—is the approximate but careful determination or judgment or probable calculation of project resources, tasks, and activities size, value, cost, and requirements. The number of people needed, the kind of skills, tasks to be completed, tasks that overlap, activities that depend on other activities to finish before they are started, costs of all resources, and others need to be carefully estimated.

- ***Scheduling***—is the process of preparing an agenda or order of project business. It is a timed plan for a project that lists events and projected operations. Good knowledge and understanding of the required project tasks, task duration, and task dependencies are necessary for effective scheduling.

- ***Organizing***—this helps with structuring work and authority relationships of members of a project group. It also helps members know and understand their individual roles and responsibilities.

- ***Directing***—this is a people-oriented function, which allows the project manager to manage the affairs, course, and actions of project members by guiding, conducting, advising, motivating, rewarding, delegating and regulating through some exercise of authority.

- ***Integrating***—this is a relatively newer function for project management. However, it is the process of bringing together and coordinating the different parts of a project system as one united whole. It is a synthesis of parts essential for the holistic functioning

of each project part, which leads to the successful completion of a project.
- **Controlling**—this may be considered the most challenging yet most important function of the project manager. It involves the process of monitoring and reporting project progress and regulating project work to conform to project goals, schedule or time, costs, scope, quality, and others.
- **Closing**—this is where every project ends. Therefore, a function helps with concluding a project. It also enables you to assess your successes and failures so that you can learn from your mistakes and, subsequently, plan for continuous or future improvement.

The Systemic World of Project Management

Possessing knowledge of all application and knowledge areas of project management can be an overwhelming responsibility. There may not be anyone who does, but understanding the principles systems is a way for you to stay on top of your project management skill. We will briefly examine the systems theory and two of its features. These features, known as the goal-seeking and equifinality characteristics of systems help us understand how we can assemble individual systems of resources to achieve one-system unprecedented goals.

The Systems Theory

The allure of the systems theory is based on the supposition that all things are relatively connected so that a change in one part of the system is a change in the whole system. This is because the attribute of wholeness fundamentally undergirds the systems concept. This concept deals with the complexities of entities, elements, and components that are distinguishable into their numbers, species, and relationships. Therefore, systems exist based on interaction, relation, and interdependence among their parts–entities and within predetermined environments.

For a relationship to exist between two or more system entities there will be constraint on the behavior of the entities involved. This constraint takes into consideration a wide variety of possible behaviors that are continuously constrained by associations to other entities in the relationship. Relationships, boundaries, constraints, and processes are fundamental features possessed by systems, and the systemic world of project management is not different.

As illustrated in figure 2.1, the project management system is a collection of objects that interrelate with one another to form a *whole*– an entire, complete, inclusive, and integrally aggregated entity. These are sets of complex elements standing in interaction while exhibiting features of causality. This means that any two or more parts of project management that are related will change due to change in any one part. To this extent, a holistic perspective to the systems' orientation provides a basis for organizing project management knowledge in terms of systems, systemic properties, and intersystem relationships.

Parts and Entities of Project Management Systems

The assortment of possibilities in developing relations and interrelations among systems leads to a pecking order—hierarchy, class, and levels of relations of parts and their connectedness. The relativity of expressions that have been used to describe interrelating parts of any system depending on their environment, ownership, and affiliation are called *subsystems, systems,* and *super systems*. Although the process of classifying and breaking down of project management systems into subsystems or even amalgamating them into super systems is really not for the purpose of unitizing them, rather it is to aid proper analysis and understanding of the world of project management.

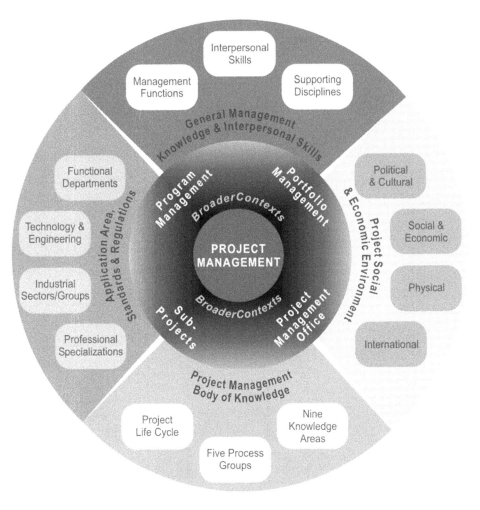

Figure 2.1 – Project Management CAKES—A composition of its Contexts, Application, and Knowledge areas, and Environment and Skills

However, the relative context under which an area of project management is considered determines how we refer to that part. Nevertheless, through a system's ability to retain wholeness and process—even when taking two parts together as minimum units—the systems characteristics of project management are not reduced. We could consider this systems feature essential in establishing that the persona of project management processes and performance is seemingly universal regardless of their application, use, or adaptation.

Project Teleology and Equifinality

To support the idea that there is no one way of accomplishing a set goal, let us consider the concept of equifinality. Theorists say it is about a goal-seeking tendency toward a characteristic final state from different initial states.

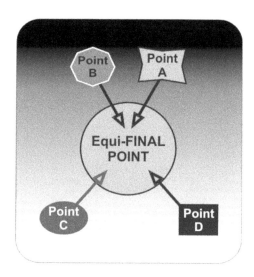

Figure 2.2 – Lines of equifinality: Illustration depicts goal-seeking lines occurring in time and from different subjective and individual (initial) states.

Goal seeking is a behavioral propensity to begin, execute, and accomplish a purpose, as this is done in the midst of complex exchanges among systems of interacting elements. Given that humans interface with these interacting organizational systems, you will find that choice plays a huge role in how goals are attained. In attaining your project goal, the processes of project management functions, such as planning and directing are, therefore, guided by choice.

Choice, is a subjective innate attribute of people, which lends itself as a function of creativity and vice versa. As a result, we choose what we do, how we do it, and when we do it. Why we do things is also a function of choice. However, equifinality helps us understand that the characteristic final state—that is, the deliverable of a project—is reachable or achieved from more ways than one or from many different starting points.

For example, the people we select as members of a project team or materials or both are chosen based on factors we consider important to the project. Arguably, the factors that inform such selection are subjective. What is essential is to have the requisite project management knowledge, so much so, that an informed selection, such that is consistent with meeting the expectations of a project's stakeholders, is done. Whatever choices you make, it is given, that the outcome of your project system will not be the same or equal to its input. This is because the systems principle is about the collective output of a system rather than the sum of its inputs.

What should be avoided is the opposite of equifinality, which is *multifinality*. A project that begins with one purpose but ends with multiple final states—deliverables—that is not consistent with the objective of the project is affected by multifinalty and, therefore, not successful. The beauty of a successful project is one that takes the different positions; knowledge, skills, and subjective goals of all involved in a project and bring them into one definite but desirable final deliverable.

Test Your PM Knowledge

6. What is management?
    a. It is the art or act of assembling human and other resources to achieve desired goals
    b. It is the process of planning, organizing, controlling, directing, and coordinating
    c. It is what managers do to accomplish goals
    d. All of the above
    e. None of the above

7. True or false: Project management is what project managers to create changes with indefinite outcomes.
    a. True
    b. False

8. One of the reasons project management is applied to a variety of business areas is because_____
    a. It provides knowledge, processes, tools, and techniques that create ways for organizing project information for success
    b. It does not help with the definition of project roles and responsibilities
    c. Options (a) and (b) are right
    d. Options (a) and (b) are wrong

9. True or false: Directing is a management function; it is also a project management function.
    a. True
    b. False

10. True or false: A project system will have parts and sub parts.
    a. True
    b. false

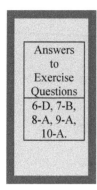

Answers to Exercise Questions
6-D, 7-B, 8-A, 9-A, 10-A.

Chapter 3

Project Management CAKES—Contexts, Application areas, Knowledge areas, Environment, & Skills

- *Project Management is not a Cakewalk, or is it?*
- *Project Management Broader Contexts*
- *Knowledge of Application Areas, Standards, and Regulations*
- *Project Management Socio-Economic Environment*
- *Project Management Body of Knowledge*
- *Interpersonal Skills and General Management Knowledge*

Project Management is not a Cakewalk, or is it?

When taken as illustrated in figure 2.1 on page 31, project management is not a cakewalk. However, if we see managing our projects as managing a sequence of connected events and activities that are conducted over a given period of time for a given purpose in order to produce a unique and definite outcome, then it is a cakewalk. This is so because managing projects should be seen as managing a tool that helps us to reach the desired end. For example, when we cut cakes during celebrations, no one really gives a thought to the table knife. Our focus is on the cake and the expectation of satisfaction we will get from taking a bite of a piece of the cake.

If you can bake a cake, remember, it is your expertise at combining and mixing those ingredients and condiments that make the cake what it is. Each cut piece of cake is essentially the same as any other cut piece of cake from the whole. However, the flavoring may differ from part to part. The project management cake is not different, either. Every system or subsystem, part or subpart of project management is necessarily undergirded by the same set of principles, processes, and procedures. In the following topics of this chapter, we will decompose and examine the project management CAKES— contexts, application areas, knowledge areas, environment, and skills, piece by piece for effective understanding of the world of project management as baked for us in figure 2.1.

Project Management Broader Contexts

The contexts within which project management exists and is currently practiced is definitively broad. According to PMBOK® Guide-

3rd ed., these include program management, portfolio management, and project management office. There are times we need to divide our projects into more manageable components. Other times we need to subcontract parts or all of our projects to external organizations or other workgroups within our organizations. Such projects are regarded as subprojects.

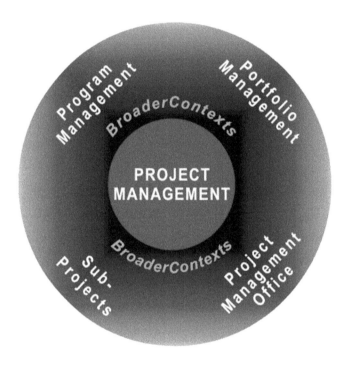

Figure 3.1 – Project Management Broader Contexts.

Subprojects

Remember our discussion on super systems, systems, and subsystems. Projects are managed under these contexts. Subprojects could be projects carried out under an aspect or a phase in the project life cycle, as part of a simple project, or as a series of events in executing a large project. Generally, their execution anchors on the same policies, procedures, and processes of project management. For example, the job of boiling water for tea can be a subproject in our tea-making project. Alternatively, the process of adding the right quantity of milk to the tea can also be a subproject.

Program Management

This is the process of using the principles of project management to manage numerously related and interdependent projects—such as those involved in building a house. It involves the integration of such projects that share a common objective. The same sets of challenges that are prevalent in managing projects are also common during program management.

Such challenges include having to deal with issues of coordinating, prioritizing, and allocating resources, managing interdependencies, and integration, and ensuring the attainment of a business objective among and across projects, departments, and workgroups. However, there are slight differences between program management and project management. This is because there are a series of repetitive undertakings involved in programs.

While a project is a one-time unique endeavor that has a definite duration, a program is ongoing and consistently implemented to achieve certain results for people or businesses. In its simplest form, let us take our house-building project, for example. Where the objective is to build a house, the whole coordinated process of planning and executing the construction of the house as separate projects is a program. Each individual process—producing the architectural and structural plans, landscaping, preparing the concrete foundations, laying the stone walls and roofing, electrical, plumbing, painting, finishing and decorating, and others—are managed as specific projects.

Portfolio Management

Large classes of projects and programs are systematically managed under this broad context of project management in order to address a business objective. It is the line of business, category of projects, the process of management, and others that determine the basis for assignment of portfolios. However, the objective of portfolio management is to carefully examine each project or program or both before inclusion in the portfolio. The idea is to maximize value and address strategic objectives.

Project Management Office (PMO)

In a business enterprise, the department or workgroup that handles by way of defining and standardizing project and program management processes is a PMO. Also referred to as a program management office, a PMO is a central organizational unit that coordinates the management of projects that are under its authority. It is responsible for project documentation and provides guidance and metrics on project management functions and execution. This is not in any way indicative that projects or programs managed by the PMO are or are not related.

PMOs add value to organizations because of their capacity to identify and develop project management methodologies and also share and coordinate project resources. They are also able to stand as platforms for developing project managers, to be a clearinghouse for project management policies and procedures, and to follow up on projects until completion.

Knowledge of Application Areas, Standards, and Regulations

The practice of project management is applicable in a variety of areas in our organizations. See the illustration in figure 3.2. As much as these areas share a common element that is important but not necessary for all projects, you are required to be knowledgeable about them.

It is also essential to be aware of the standards and regulations that apply to each application area, especially as they relate to your project.

> *Regulations* are legal restrictions promulgated and imposed by a government authority, which specifies certain characteristics of a service, process, or product.

According to PMBOK® Guide-3rd ed., these areas are defined in terms of (a) the ***functional*** responsibilities of various departments and applicable disciplines in our organizations. These could range from the production to human resources, logistics to engineering, IT to accounting, legal to marketing, and other departments; (b) specific kinds of technology and engineering, such as software development, or construction, engineering, and others. Such comprise the ***technical elements*** or processes, which we need to be aware of or be knowledgeable about; (c) ***specialty management*** knowledge of operations in specific areas, such as government contracting, community development, and financial services contribute to the successes of our projects; and (d) being knowledgeable of ***industry***

sectors or groups aid in helping us keep focus on project objectives. Such industry groups are IT, financial services, aerospace, automotive, agricultural, chemical, and others.

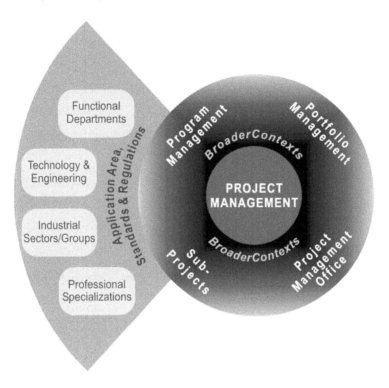

Figure 3.2 – Project Management Application Areas, Standards and Regulations

Standards and regulations are norms and mandatory requirements established by recognized professional bodies and governments, respectively. Usually, standards are formal documents that are established and approved by consensus to address uniform rules, criteria, methods, processes, and practices in technical and other fields. Whereas regulations are government-imposed, promulgated requirements address provisions for mandatory compliance with specifications relative to certain characteristics of a service, process, or product.

It is fundamental that you or at least a member of your project team is aware and knowledgeable of existing professional and industry standards, for example, such that address calibration, metrology, quality, safety, compatibility, performance, and other standards issues. In the same vein, it is of high importance to have the knowledge or at least have access to knowledge of government regulations in domestic and international environments. For example, it is important to be aware of regulations on prices and pollution, wages and employment, emission and industries, importation and exportation, and others.

Ignorance of these kinds of government regulations can mar the success of a project both in domestic and international environments. It is necessary that you or a member of your project team has adequate knowledge or is at least aware of applicable government regulations. Ignorance of such impacting environments can lead to project failure.

> ***Standards*** are formal documents established and approved by consensus to address uniform rules, criteria, methods, processes, and practices in professional fields.

Project Management Socio-Economic Environment

A project management environment includes the external conditions and resources with which it has to interact. You cannot avoid or evade the cultural, social, and economic environmental contexts of your project. See figure 3.3. This is because all projects begin and end in these environments. Every project impacts and is impacted by its environment. Therefore, you must be aware of the environmental forces that can make or mar your project.

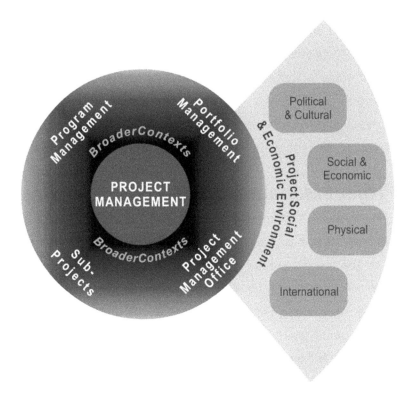

Figure 3.3 – Project Management Social, Cultural, Economic, Political, Physical, and International Environment.

All projects are planned and implemented within the context of their peculiar environment. So you need to know how your project would affect or be affected by its *cultural and social environment*. This does not exclude the culture of your organization, which should be equally examined.

Your goal should be to seek to understand all or parts of what is important to the people your project would affect or that would influence your project. This should include understanding their interests, economics, and demographics, ethnic, religious, and educational characteristics. Equally important is learning about the projects' *political,* as well as its *international environment*. Considerable factors are the local, national, and international customs, mannerisms, and climate. You should also note that having knowledge of the local geography and ecology relative to the project keeps you abreast of the requirements of the project's *physical environment.*

Project Management Body of Knowledge

The body of knowledge of project management is the core of this profession. It is comprised of three main sections—the project life cycle, the process groups, and the knowledge areas. The second and third divisions of this book are dedicated to discussing the project management body of knowledge in detail. As one who seeks project management knowledge, your having adequate understanding of these three areas will ensure a competent practice of project management. They are the cornerstones that undergird the practice of project management. See illustration in figure 3.4.

While the project management body of knowledge remains unique to the discipline of project management, some areas share common characteristics with other management fields. It is, therefore, necessary that you should have or at least acquire a certain level of skill and knowledge of general management principles.

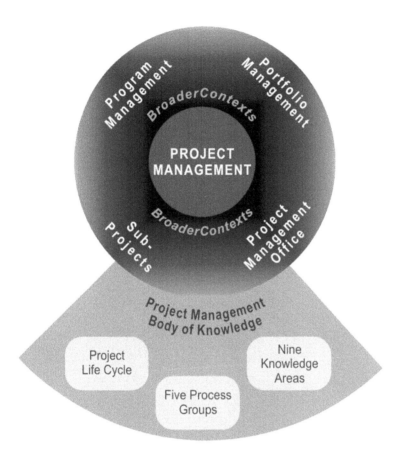

Figure 3.4 – Project Management Body of Knowledge.

Interpersonal Skills and General Management Knowledge

We have already established that management is an art and act of managing human and other resources in order to accomplish the desired goal. This is achieved through performing functions, such as planning, organizing, directing, controlling, and coordinating. For the field of project management, additional functions such as scoping, estimating, scheduling, executing, and closing are also performed. However, general management functions and skills are enhanced by knowledge of supporting disciplines. See figure 3.5.

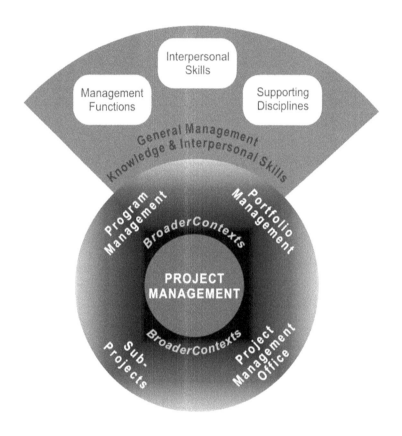

Figure 3.5 – General management knowledge and interpersonal skills of project management.

These disciplines range from strategic to tactical planning, operations to manufacturing and distribution, and from purchasing to sales and marketing. Others are accounting and financial management, contracts and law, personnel management and organizational behavior, logistics and information technology, and others. To possess knowledge and experience in some of these and more areas lay the foundation that helps to develop skills that are essential for successful project management.

> *Your Role as Project Manager* includes the following:
> 1. Utilize available resources to produce deliverables within tine, cost, scope, quality, and other performance constraints.
> 2. Act as interface between customer and organization and manage conflict resolution effectively.
> 3. Make necessary and required decisions.
> 4. Negotiate with stakeholders to accomplish project tasks.

The **development of core competencies** in project management is also a function of your interpersonal skills. It is recommended that for effective interpersonal relationships, you should possess and utilize a set of skills that are premised around your role as a project manager and your ability to

- Be committed to achieving business goals while maintaining relationships and quality throughout your organization,

- Influence people by recognizing behaviors and reacting appropriately and respectfully to them, understand and use organizational policies and politics well, anticipate the impact of project decisions, and seek and obtain cooperation from all stakeholders,

- Manage people competently through coaching, directing, delegating, motivating, training, and developing; communicate

effectively by listening, monitor and control plans, schedule, budgets, work progress, and reward appropriately and adequately.
- Solve problems analytically, be creative and diligent, take calculated risks, obtain factual information, and think conceptually,
- Manage self, have integrity, be confident and consistent, truthful and honest, timely and orderly, trustworthy and tolerant, firm yet flexible, and able to work under pressure,
- Show and manifest wisdom in responding to project situations and accepting responsibility for every aspect of the project.

Although the foregoing traits (listed in table 3.1) are not exhaustive, they establish an environment that should help you focus on or pay attention to what is important. Essentially, you should know **your field** of operation, have excellent **communication skills**, possess appropriate and adequate **organizational skills**, and exhibit good **leadership skills**. It is also advisable to be sensitive to morals and ethical issues.

It should be understood that while some of these competencies are learnable, others are acquirable from business and life experiences. Some have argued that people are born with certain traits like those of leadership and so are not learnable. Others have argued that management traits are completely different from leadership traits. That may or may not be true; however, our environments contribute a great deal to what and what not we become.

To this end, being in the right environment can expose you to a set of balanced skills you need to lead and manage projects successfully. It is your responsibility to lead and manage a project to its success. In this regard, it is the specific project assigned to you as a project team member or manager. It is an awesome responsibility to be both a leader and a manager; therefore, you must possess a mindset that will enable you to perform as such.

Table 3.1

Leadership and Management Traits:

Leaders	Managers
try to change organizations	sustain and control them
take risks especially when payoff is high	are cautions and take fewer risks
have visions of how an organization can be better	follow present vision
use power for influence	use power to coerce people into compliance
use conflict to identify options for action	will rather avoid conflict
have risk-taking behavior that can plunge an organization into chaos	try to solve problems to bring order while building commitment to organizational goals
are important change agents so play key roles in fast changing external environments	play key roles in stable external environment

There is a need for you to be creative, balanced, flexible, and responsive to project needs. You have to be comfortable with change, uncertainty, and taking of calculated risks and be technically and financially knowledgeable. You should be entrepreneurial and productive, systemic and process oriented in your approach to managing projects, and be culturally and disciplinarily diverse.

Test Your PM Knowledge

11. Which of the following is not part of the project management system?
    a. Contexts and knowledge application areas
    b. Knowledge areas and environment
    c. Skills, standards, and regulations
    d. Performance and international environment

12. True or false: Program management is similar to project management because it uses the principles of project management to manage numerously related and interdependent projects.
    a. True
    b. False

13. In your organization, the department or workgroup that should handle standardization and definition of project and program management processes is a _____
    a. PPMO
    b. PPM
    c. PMO
    d. OPPM

14. Is it true that being knowledgeable of industry sectors help in keeping project managers focused on project objectives?
    a. Yes, it is true
    b. No, it is not true
    c. Maybe it is true
    d. May be it is not true

15. Which of the following is not part of the project management socio-economic environment?
    a. Social and economic
    b. Political and cultural
    c. Information and technology
    d. Physical and international

Answers to Exercise Questions
11-D, 12-A, 13-C, 14-A, 15-C.

Division 2

Project Management Cornerstones

Chapter 4

Project Life Cycle, Phases, and Process Groups

- *Project Life Cycle*
- *Phases and Stages of a Project*
- *Life Cycle Characteristics*
- *Process Groups – Initiating, Planning, Executing, Monitoring and Controlling, Closing*

Project Life Cycle

A projects' life cycle defines the beginning and the end of that project. We can see the series of developments and changes that a project undergoes from its beginning stages to the reoccurrence of the same stage in another project. Although these stages vary by industry and project characteristics, organization, and application, they, however, define the nature of work done in each stage and the skill required to do them within the stage. This is comparable to the life cycle of a living organism that begins at birth, goes through development, matures, begins to age and depreciate, then dies. If this is taken within the project management context, we will have three phases of a project life cycle—the initial, intermediate, and final.

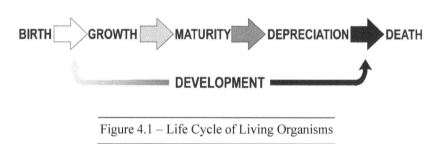

Figure 4.1 – Life Cycle of Living Organisms

Each of these phases roughly defines, clarifies, and connects the beginning of a project to its end. The point is that it helps you clarify the inputs and outcomes of one phase as you connect to the next. Although most project life cycles share common characteristics, transitioning from one phase to another requires some form of review for completeness and accuracy and maybe approval before starting on the next phase. However, there is no one general way of defining a project life cycle for any particular project or industry.

Project managers in different organizations and in different industries define and choose what the ideal life cycle suitable to meet their goals is. According to PMBOK® Guide-3rd ed., however, there are common practices within specific industries that lead to the use of a preferred project life cycle. These common project life cycles generally define

1. The kind of technical work to accomplish in each phase.
2. When and what kinds of deliverables each phase is expected to generate, how it should be delivered, reviewed, and or validated.
3. Who and what is involved in each phase.
4. How each phase is controlled and approved.

Industry Preferred Project Life Cycles

The following describes some preferred industry-specific life cycle practices. *SDLC* is an acronym for Systems Development Life Cycle, which is employed in the information technology industry to describe a variety of systems development processes and phases. A simplified, yet classical SLDC has four stages that must be completed for any systems development project—project initiation, system analysis, system design, and system implementation. A more contemporary project management systems development process has eight phases—scope definition, problem analysis, requirement analysis, logical design, decision analysis, physical design and integration, construction and testing, and installation and delivery.

In new product development, there are seven lifecycle stages—ideas generation, ideas screening, concept development, business analysis, prototype and market testing, implementation, and commercialization. Similarly, for a macro project, the engineering industry would use the following life cycle stages—ideation, concept design, detailed design, tool design, production planning, and manufacture.

It is also distinctive in other industries. In the marketing and sales industry, the lifecycle starts from idea generation and proceeds through advertising and selling and ends at sales order processing. For each unique lifecycle application, it is merely about knowing how to deal with managing descriptions and elements of a project through its conception, development, implementation, and conclusion. From a business and maybe an industry point of view, these four elements are common denominators that describe a typical life cycle.

> *Generic Project Phases.*
> The Initiation Phase
> The Planning Phase
> The Execution Phase
> The Closing Phase

Phases and Stages of a Project

Regardless of the industries' project life cycle preference, a generic project life cycle may possess and be defined according to the following four phases:

1. The **Concept or Initiation** phase—identifies and defines the problem to be solved. At this stage, the problem may be unclear, but feasibility studies may be consulted to bring some clarity to the problem. Based on this clarity, a proper goal and objective of the project can be defined at this point. You can also identify who your project stakeholders and project team are. You should be able to answer questions on the risks involved and their alternatives, as well as obtain initial cost estimates. The output of this phase is the project charter or some kind of contract—approval—that permits you to proceed to the next phase.

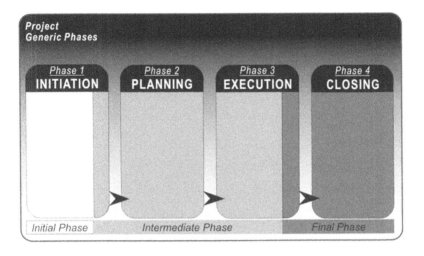

Figure 4.2 – Generic Phases of a Project

2. The **Planning or Development** Phase—determines what must or needs to be done by developing a problem statement that is consistent with the goals and objectives of the project. You may have heard it said that to fail to plan is to plan to fail. This phase of any project could be regarded as the most important. It is where you develop a blueprint for the project. The success or failure of your project can begin here. In this phase, you develop strategies and a detailed work scheme for achieving your project goals and objectives. Your step-wise plan should include defining, determining, and establishing the scope baseline, budgetary costs, a risk assessment, activity definition and work breakdown

structure, human resource requirements, tasks, and responsibilities, authority levels, control systems, quality standards, communications patterns, and procurement schedules.

3. The ***Execution and Implementation*** Phase—directs and manages the project execution by establishing clear and effective lines of communication between stakeholders and the project team and between teams and team members. Within this phase, you can also perform quality assurance, acquire and develop project teams, determine definitive estimates, procure goods and services, resolve problems and conflict issues, and monitor and control scope, time, cost, and quality.

4. ***Closing and Termination***—prepares the project for review, handover, and acceptance. You will finalize the project by evaluating and documenting results and lessons learned for future use, release or redirect, or reassign project team and other resources, while project or product responsibility is transferred to the project owner.

Project Management Phases and Common Characteristics

We have learned from the foregoing that activities that occur in each project phase can be very detailed. While project management phases remain generally sequential, they do share some common characteristics. The type of technical information they transfer from phase to phase defines these characteristics. You need to commit this type of technical information to memory. This is because such information helps you stay on top of your game as a project manager, especially when you make plans, communicate with stakeholders, source for human resources, negotiate for resources, and even resolve project conflicts. For example, knowing the amount and level of staffing costs at all phases of the project will help with proper budget and human resource management.

You should also know when a stakeholder's ability to influence a project is highest or lowest, or even the riskiest or most uncertain part of the project throughout its life. Knowing all these and more can help reduce the ifs, buts, assumptions, and constraints information gap regarding your project. We will graphically consider the different scenarios of these characteristics and factors, so you can bridge the information gap when managing your projects.

The following figures 4.3, 4.4, 4.5, 4.6, 4.7, and 4.8 were adapted under permission from Project Management Institute *A Guide to the project management body of knowledge (PMBOK® Guide) –* Fourth Edition, Project Management Institute, Inc. 2008, figures 2-1, 2-2, and 3-2.

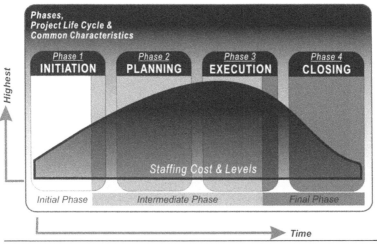

Figure 4.3 – Project Common Characteristics: Staffing and Cost Levels

Staffing Cost and Levels—are low at the beginning phase, highest at the intermediate phase, and drops rapidly at the final or closing phase of the project. This picture becomes clearer knowing that staffing costs involve the costs of finding the right people—recruiting. That is people with the right skills and abilities that will fit into your project team or organization. You could also consider costs of project functions, such as staff placement, appraisal, rewarding, evaluation, and general human resources functions as costs of staffing.

Project Risk and Uncertainty—is highest at the beginning, because the probability of successful completion of the project is lowest at the start of the project. So much so, that if the project is properly planned, there are no disproportionate inaccuracies in cost estimates and schedules, and the project is adequately and appropriately staffed.

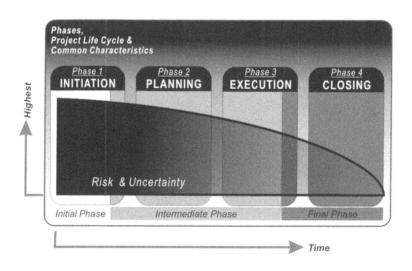

Figure 4.4 – Project Common Characteristics: Risk and Uncertainty

However, as specific knowledge and accurate information required for the project are increasingly acquired as the project progresses, sources of risks—ambiguities, assumptions, and guesstimates—are eliminated. Predictions then become more accurate and, therefore, uncertainty and risk are increasingly reduced as the project winds down towards its final phase.

Stakeholders' Ability to Influence—is fundamental for every project. These people are or can be affected by your actions and the deliverable of your project. Therefore, they possess ultimate power, influence, and say-so over your project at the beginning. Consequently, their influence over the outcome of the project is highest at the start, but it progressively declines towards the final phases of the project.

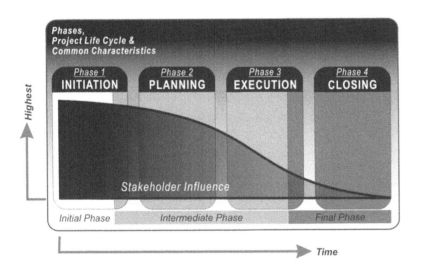

Figure 4.5 – Project Common Characteristics: Stakeholder Ability to Influence

Cost, Resources, and Amount at Stake—in terms of financial exposure and resources to be and already committed to the project are lowest at the start of the project. These increase as the project progresses. How resources are controlled is important to the success of a project. Resources help add value to a project; therefore, their value depends on the level of need we attach to them, how they will be used, and how they will change over time. As a result, knowing that project cost and resource utility increases as the project progresses and are highest at the end of the project help with effectively managing successful projects.

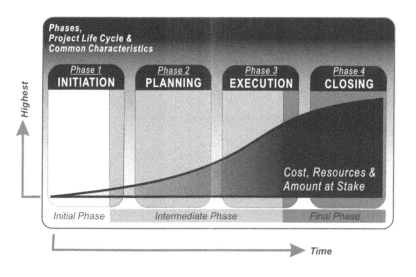

Figure 4.6 – Project Common Characteristics: Cost and Amount at Stake

Cost of Changes—attributable to changes we make to project scope, schedule, or budget are lowest at the start of the project. Such costs include direct and indirect expenses we have or will incur in acquiring more materials or labor or both for effecting additional changes to the project. These kinds of changes are necessary to satisfy the objectives of the project and meet stakeholder expectations. The cost of changes increases with time as the project progresses, and it is at the highest in the final phase of the project.

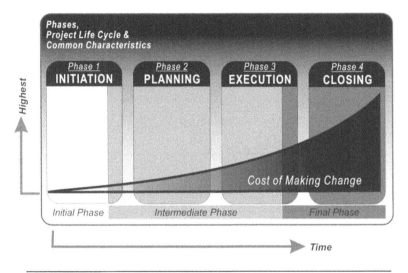

Figure 4.7 – Project Common Characteristics: Cost of Making Change

Value Added—represents the process of combining, using up, and converting tangible and intangible resources to create a project deliverable. A project has to have value for its stakeholders; therefore, we can calculate value-added as the difference between a project's final value and the direct and indirect cost of doing the project. Consequently, the opportunity to add value to any project is highest at the beginning and some parts of the intermediate phases of the project. As the project winds down, the opportunity to add value to a project also decreases. Therefore, if your project must deliver value to its stakeholders, how you manage the early phases of the project is very important to accomplishing value for the project.

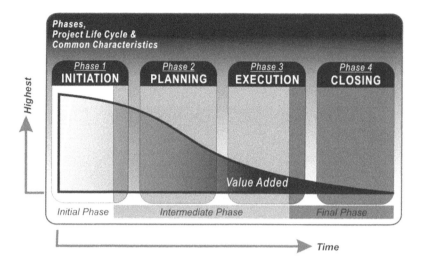

Figure 4.8 – Project Common Characteristics: Value Added Opportunity

Process Groups–*Initiating, Planning, Executing, Monitoring and Controlling, Closing*

In discussing project management phases, we have laid the foundation for these five process groups. While it may appear repetitive and confusing, they are completely distinctive. According to PMBOK® Guide-3rd ed., project management process groups are not the same as its phases. The reason is that project phases are dependent on industry, practitioner, and application preferences. Therefore, phases vary and can include more or lesser levels or may even be labeled differently. For example, in information technology, a project phase is described according to how large or complex that project is or what objective the project life cycle seeks to accomplish.

If you were to manage a newly developed computer software release project, you would adopt the following life cycle phases—pre-alpha, alpha, beta, release candidate, and gold. To people who are not information technology or computer software engineers, this is a foreign language. However, it respectively means to (a) issue a non feature-complete software before testing, after it has been built; (b) deliver software to testers; (c) release software outside the organization that built it for real-world evaluation and technical preview; (d) check software, which has potential as final product and ready to be released for emerged bugs or codes completeness; and (e) release the production or stable live version software.

> ***Equipment Design and Development Phases:***
> 1. Conceptualization
> 2. Concept development
> 3. Design
> 4. Prototype construction
> 5. Testing

In the same manner, the project phases of simple equipment design and development would be described as—conceptualization, concept development, design, prototype construction, testing. Process groups are the same regardless of industry, preference, or application. They do not change; therefore, they are repeated for each phase of a project or subproject. The point is that these processes are present in each phase of and throughout the project. Remember that in each

> ***Computer Software Release Phases:***
> 1. Pre-Alpha
> 2. Alpha
> 3. Beta
> 4. Release candidate
> 5. Gold

phase of a project, certain activities are started, developed, and concluded before transitioning to the next phase.

For example, in the planning phase of a project where you would develop a detailed step-by-step scheme, you would go through processes of initiating and planning the development of the plan. This is followed by executing the plan development, monitor and control relevant steps to ensure that your plan development is consistent with the objectives of the project.

Subsequently, the plan development will end by reviewing, formalizing, and accepting the results of this phase in order to transition to the next phase. The underlying and guiding principle here should be to plan it, do it, check what you have done, and then take action. Note that many projects fail because they are transitioned from phase to phase without appropriately following these processes.

What is a Process?

It is appropriate at this point to examine the meaning of process. In many business organizations, process as a term is not new especially when related to improvement. However, not many possess adequate understanding of it or effectively manage and utilize it to accomplish business objectives. According to *Webster's New World Dictionary & Thesaurus* (2005), a *process* is generally defined as a particular method of doing something that involves a number of steps.

The simplicity of this definition is not sufficient to convey the dimensions of operations of 'process'. Nevertheless, the significance of its description is that it is a scheme that leads to the accomplishment of something. In many organizations and in the many things we do, process levels exist. Therefore, they perceived to be frequently rolling or stumbling along whether we attended to them or not. Process(es) exist in multi-levels that present opportunities for examination of an input-output linkage depicted in a process map.

Given that between every input and output is a process, our understanding of process would be incomplete if we do not examine the processes through which inputs are converted to outputs. The implication is that processes exist within processes that we may be ignorant of. This indication of abstraction paints process in a way that permits it to be viewed as abstract, therefore, construed and constructed in many forms.

To make it real, we need to model process in some way that it is to be understood, and importantly, in roughly the same way by two or more people. Consequently, process has been defined by experts and theorists, especially from a business or organizational point of view as (a) a set of interrelated activities, (b) a series of steps designed to produce or receive a product or service, (c) a value chain based on its contribution to the creation and delivery of a product, and (d) a set of related (and interdependent) formal and/or informal activities that produce specific end products.

While these definitions implicitly or explicitly preclude the human facilitators of process, they are insufficient to convey a comprehensive meaning of a process. Given the importance of people in facilitating processes, whether in planning, executing, or managing the series of steps involved in project management processes, it is essential to include them in the definition of a process.

Time, as a factor, is also important, because these steps of interrelated activities need to occur in a given period; otherwise, there is absolutely no sense in having a perpetual process. The use of tools, techniques, and technology in facilitating process cannot be overemphasized. Whether in the form of a chart, or technique, an appliance or equipment, and machinery or automation, project tools and technology are essential in enhancing the speed, accuracy, and efficiency of your project work.

Consequently, 'process' is defined as a means to accomplish a predetermined goal, which exists in a wide range of relationships among time, technology, and people who facilitate it. In other words, it is a function of time, technology, and people.

The Five Process Groups

The five project process groups are the initiation, planning, and executing, monitoring and controlling, and closing process groups. They are connected by their outcomes or deliverables, as the outcome of one process group is the input to another. You should also know that contrary to the sequential connection between the project phases, the process groups are not strictly connected or related linearly. However, their relationship exhibits a general flow from initiation to planning to executing and to closing. The monitoring and controlling process group are linked to all the process groups.

> *The Planning Process Group* is the most significant and all embracing of all process groups given its level of interactiveness, iterativeness, and capability to bridge dreams and reality.

Initiating

Within this process group, you will define and authorize the start of a new project or a new phase. For the most part, this process is usually not within the boundaries of the project scope of control. At the project level, you need to be aware of the environmental factors that will influence your project. They could include your organization's culture, information system, policies and procedures, standards and guidelines, processes, historical information, and the human resource practices. However, at the project phase level, you could also consider the results and lessons learned from working on previous phases. These make up the input to this process group. The outputs, which should feed into the next or other process groups, are the project charter and the preliminary scope statement.

Planning

The planning process group is by far the most significant, the most extensive, and the most interactive component of all process groups. It is the function that bridges dreams and reality, connects concepts to configurations, and produces a program for execution. You would begin performing this function by defining and refining the project objectives and preparing and laying a course of action that should lead to attaining the objectives.

The success of the project would anchor on how well you are able to gather and use information from internal and external, as well as other sources to develop a detailed plan. This detailed plan is used to guide project execution, document assumptions and decisions, facilitate communication, and define project reviews. In addition, the plan helps provide baseline for measuring progress and performance, control the project, and establish a management strategy.

Planning also extends to defining and planning the project scope, creating a work breakdown structure, defining and planning project activities, activity resource, activity sequencing, activity estimating, and project schedules. Cost estimating, quality, human resource, communications, and risk management, and procurement planning is all part of the planning process group.

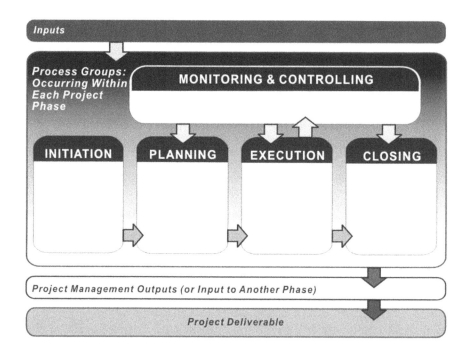

Figure 4.9 – Project Process Groups

Executing

The detailed plan of action is implemented under this process group by acquiring, integrating, and utilizing human and other resources to accomplish the project objectives. You would also develop and manage your project team, assure quality performance, and communicate and distribute information to stakeholders, suppliers, and vendors. In instances where there are scope and other change requests, they are handled under this process group.

Monitoring and Controlling

This is simply a check and control process. The criteria established during initiation and planning provide the yardstick for identifying variances based on the plan, and for monitoring project progress. In non-conforming situations, you would take corrective actions in a timely manner to control project execution. Monitoring and control also include scope verification and control, schedule, cost and quality control, risk monitoring and control, and ensuring the management of stakeholder communications. You would also manage review, document, and manage the contractual relationship with your project suppliers and vendors.

> The check and balance capability of the *Monitoring and Control* process group make it a fundamental component that help ensure deliverable of intended project quality.

Closing

The project ends under this process group. You would verify the completion of all defined processes within other process groups and formally prepare project for handover or transfer to other project phases.

Interaction of Process Groups

In as much as all five processes begin at a point in the project, they also end at a point during the project. Their execution is not distinctive or isolated; rather, they overlap and interact with each other, within a phase or a project. However, the division of projects into phases compels the process groups to drive project performance effectively from beginning to completion in a controlled manner.

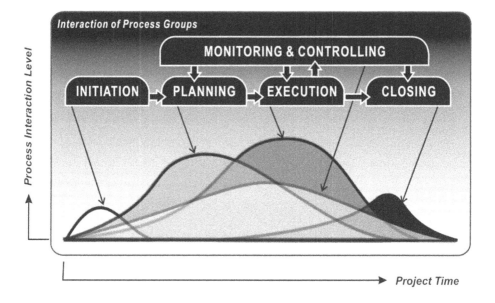

Figure 4.10 – Project process groups interaction within a phase or a project. Adapted under permission from Project Management Institute *A guide to the project management body of knowledge (PMBOK® Guide)* – Fourth Edition, Project Management Institute, Inc., 2008, figure 3-2.

This means that the processes are repeated within each phase, especially for projects with multiple phases, until the objective or criteria for completing each phase is met. In the interaction of all processes, the initiation and closing processes rarely meet. However, the planning, execution, and monitoring and controlling processes overlap and interact with all five-process groups. Figure 4.10 shows how the process groups interact within a phase or project.

Test Your PM Knowledge

16. Each project phase defines, clarifies, and connects the _____ of a project to its _____
    a. Beginning, middle
    b. Beginning, end
    c. Initiation, development
    d. Execution, closing

17. The generic project phases include the following phases
    a. Ideation, design, tooling, and production
    b. Concept, analysis, testing, and implementation
    c. Initiation, planning, execution, and closing
    d. Concept, implementation, development, and manufacture

18. At what phase of a project is project uncertainty highest?
    a. Initiation
    b. Planning
    c. Implementation
    d. Termination

19. Your project stakeholders' requests that specific changes be made to your project. At what stage do you think it will cost less to make any kind of change?
    a. Initiation
    b. Planning
    c. Implementation
    d. Termination

20. How many process groups are there in a project?
    a. two
    b. three
    c. four
    d. five

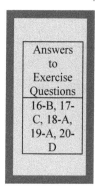

Answers to Exercise Questions
16-B, 17-C, 18-A, 19-A, 20-D

Chapter 5

Managing Small and Simple, Large and Complex Projects

- *Small and Simple Projects*
- *Manage Small and Simple Projects Using Project Phases and Process Groups*
- *Large and Complex Projects*

Type of Project Distinction

In terms of the principles of project management that guide project execution there is relatively little or no difference between small and simple, large, or complex projects. Application and environment may differ, but a project is a project whether regardless of size or complexity. Nonetheless, the simplicity of a project is premised around the state of composition and singleness of the project. On the contrary, a complex project is one that exhibits a labyrinth and levels of interacting project activities. In the same vein, relative size and extent delineate a small or large project.

Small and Simple Projects

Our tea making, book reading, and house building projects may be categorized as small and simple projects. However, the relative size of the deliverable, in addition to the relative simplicity of the tasks involved and performed by a relatively small number of people can keep these projects classified as small or simple. However, other factors contribute to why projects are categorized as small or simple.

They are relative costs, project duration, skill areas, number of people involved in the project, clear definition, and straightforward objective. Others are a narrowly defined project scope, require

---

Three main **Factors** that **Distinguish Small** and **Simple** from **Large** and **Complex Projects**:
- Fewness of skill areas
- Broadness of scope, and
- Complexity of deliverable

relatively little or no automation, and produce a clear and straightforward deliverable.

For more apparent distinction, the interplay among three factors separates small projects from large and complex ones. One factor is the level of complexity of the project deliverable, the other is how broad the scope is, and the last is the number of skill areas involved. For example:

- A project that involves a few skill areas but produces a complex deliverable is not small.
- A project that involves a few skill areas but has a broad scope is not small.

The explanation to the above is that a project with a broad scope will definitely involve the use of more effort; therefore, more skill areas as these would result in deliverables that are more complex. In addition, the levels of management or supervisory authority, which is a function of effectively managing a variety of skills, also distinguish a small project from a large one.

The **Difference between Small and Simple Projects**:
- Shorter Duration
- Involves from one to three people
- Lesser project management formalities

In further distinguishing a small from a simple project, a *simple project* is one that uses even lesser skills, has a lesser scope, and produces a far simpler and more straightforward deliverable. For many, this could amount to performing an assignment, especially when such assignments possess the elements of a project, such as having a definite beginning and an end, are temporary and have unique outcomes. The main difference between a small and simple project is that simple projects are completed within shorter time periods, involves one and up to three people and use less of the formalities employed in typically larger and more complex projects.

Managing Small and Simple Projects

The power of project management is its flexibility to be used for all categories of projects. More importantly is the opportunity it affords you to clearly define your expectations, use resources better, and minimize the disappointment and dissatisfaction from wasting time and effort. There is value in using project management knowledge in managing your personal or professional projects or both. You would use project management processes and tools to enforce a disciplined attitude for organizing your project data.

It would also provide a means to define your project goals, scope, roles, and responsibilities so that you can focus on priorities while you effectively manage time, cost, and scope in order to deliver stakeholders' expectations.

Regardless of the size and scale of your project, there are challenges you must address. These challenges are no different from what you would do if it were a large project. As we have established, you will plan your project well, so that you can adequately define your scope, cost, and time required to finish the project. This is because the plan will help you determine how to keep your project stakeholders informed and control the project, which increases the chances for success.

You may also encounter challenges that border on how important your small project is to your organization, available resources, and appropriate skills and the right experience of team members. Others include lack of knowledge in specific areas, managing multiple smaller projects within a project, and access to the right tools and processes.

> A **Project Champion** is an enthusiast for a project who convinces people to get them to see the importance of a project. Thereby, working with team members and stakeholder to accomplish the goals of the project.

Such challenges are not meant to discourage you. Rather, they are opportunities for you to rise to the challenge by becoming a ***project champion*** for the project. In addition, you would also become a skills

Table 5.1
Features of Small and Simple Projects:

Small Projects	Simple Projects
Involve 10 or lesser team members	Involve 3 or lesser team members
Involve a few number of skill areas	Involve fewer number of skill areas
Employ more project management formalities	Employ lesser project management formalities
Short in duration	Shorter in duration
Less broad or narrowly defined scope	Straightforward scope
Can be for a single department or part of a large project	Part of a small project or equivalent to an assignment
Costs are small	Costs are smaller

developer, a subject matter expert, a facilitator, a multi-tasker, and effective chooser of the right project management tools and processes. Combining all these skills and more lead to the success of your project.

Managing a Project Using Project Phases and Process Groups

Let us consider managing a house-building project as a simple and as a small project using the phases and processes of a project. As a simple project, you could be asked to construct the paper model of the house, but as a small project, you are in charge of managing the construction of one of the houses in a housing scheme program. Remember that the phases and processes for either type of project will be the same. The only difference is the scope, the amount of skill required, and the duration, the cost, and others.

Only you or two others, in addition, can complete constructing the paper model of the house in a shorter time than it will take to build. As a result, the cost is lesser; the amount of skill fewer, scope is narrower, and the deliverable would be far less complicated. Compare managing the construction of 100 units in a housing scheme to managing a single housing unit. This is a small project to manage.

Construct a House Model and Build the House

Let us take a pragmatic approach to acquiring an adequate understanding of the principles of project management. This is the scenario. You are part of a community development organization, and you are assigned the project to construct a scaled-down paper model of a four-room, two and half bath, housing unit. Thereafter, you would manage the project of building the house. Given the relative size, scope, and duration of these projects, they are by no means simple and small projects. So how do we utilize the principles of project management to manage this project?

> A **Project Charter** is a document that a project Sponsor or Initiator issues to authorize a project formally. This document also gives a project manager the authority to apply organizational resources to execute a project.

We have determined that projects in the simple project category do not use all the formalities of project management. However, those in the small projects category do. Therefore, we will consider some simple project management tools and methodologies. The use of these tools helps ensure effectiveness in managing projects in these categories. So let us begin by determining what we should in the processes of initiating, planning, executing, monitoring, and closing the project.

Initiate the project. The first step is to get authorization for this project. This is usually done by developing a project charter. The project charter describes the objective, the scope, the duration, the budget, and such other information relevant to starting your project. For small projects, the project charter is usually more detailed, and the following would apply more to small projects than simple ones.

1. Develop the Project Charter—you could develop a project charter to establish clarity of project objectives and expectations and set the stage for the planning phase. For a simple project, or if it were a personal project, a project charter may be too formal. You could then simply establish the start-up needs of your project. Do so by

a. *Defining, Describing, and Stating* the purpose of the project, which should help you perform the activities you need to start your project. Your descriptions, as basic as it may be, should include **what** you want to achieve, **why** you want to achieve it, **when** it should be achieved, **how** it should be achieved, and **who** should achieve it.

b. *Consider the Inputs* to this process; this is to utilize pre-project information, contract, and statement of work, organizational processes, or environmental conditions to develop the charter. contract, statement, and historical information.

c. *Develop the Output* from using the rough descriptions of the project objectives, the input considerations, and talking with sponsors and stakeholders of the project. The output is a project charter.

2. Include this Information—note that different organizations use different project charter formats. However, most formats will include the following information:

- the roles and responsibilities of project sponsors, managers, stakeholders, and team members
- the description of the project objective, background, boundaries (that is within outside scope), duration, and budget
- information about the project that lists the deliverable, the assumptions, the constraints, dependencies, risks, and opportunities, and
- supporting information, such as the project's impact on a new or existing business process, as well as the stakeholder's criteria for project acceptance. See Chapter 19 for a sample of a project charter.

A **WBS structure** levels can begin from the Program level through the Project to the Phase, the Task, the Subtask, and finally the work Package. However, a work package can further be broken down into more manageable components, known as activities.

Plan your project. Next, you would develop a plan using simpler versions of the tools used for large projects so that it can help you stay focused on the project objective. Begin by reviewing the project objective or the project charter and such other documents that provide background information on the project.

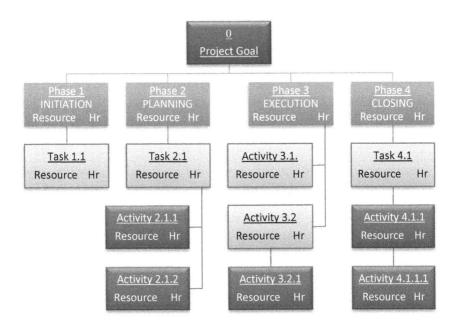

Figure 5.1 – WBS, decomposition of phases, tasks, activity, resources, and duration (Hr).

Speak with or interview project stakeholders to gather more information about the project. Then develop a plan that decomposes the project deliverable into smaller tasks and activities.

Tools to use include a to-do-list (especially for a simple project), a simple WBS—work break down structure (figure 5.1), or an integrated resource chart and task plan (figure 5.2). Graphical Planning—the WBS presented below is a basic version, but a good tool that helps show a considerable level of hierarchical detail used to execute projects. It is used to decompose or break down project activities into manageable units from top to bottom. The first level is used to list the project phases (for a simple project you could skip the phase level), the second lists the tasks, and the third lists the more detailed tasks. You can also include names of resource persons and the number of hours for each activity and task in each bloc as some practitioners do.

1. Integrated Resource Chart—using the WBS, we can proceed to develop a more integrated action plan. The chart lists the tasks and activities using a military-style numbering scheme to present the decomposed project. The chart also provides columns for listing resource persons and estimated start and finish dates for each task and activity. There are also two columns for showing project status and how tasks overlap; that is, how separate tasks can be performed at the same time. This is a horizontal shaded bar—Gantt chart that depicts project tasks against project time or duration and progress.

Execute the project. Now that we have a detailed plan, we know what task and activity to perform, when to start and finish it, and how they overlap with other tasks. Focus should now shift to executing and making sure that the implementation is consistent with the project goals and plan. The goal of project execution is to apply, deploy, and manage resources to accomplish the project deliverable. Remember that good project teams make things happen, so in managing your small team, share the project objectives. Select your team based on their skills and ability to work in a team, contribute to building the team, have strong potential for solving problems, and share the successes and failures.

Integrated Resource Chart and Plan for Managing Simple Projects							
Project Name:					Project Owner:		
Project Manager:			Prj/No:		Type: Date:		
No.	WBS – Phases /Tasks / Activity	Human Resource	Start Date	Finish Date	Duration / Progress Bar (hrs) 8 · 16 · 24 · 32 · 40		Status
1	Initiation						
1.1					■■■■		CP
2	Planning						
2.1					■■■□□		IP
2.1.1					■■□		IP
2.1.2							
3	Execution						
3.1					■■□□		IP
3.1.1					■■□		IP
3.1.2					■□		IP
3.1.3.1					□		NS
4	Closing						
4.1					□		NS
4.1.1					□		NS
4.1.1.1					□		NS
					PP		

Figure 5.2 – Integrated Resources Chart and Plan
Key: CP=Complete, IP=In Progress, NS=Not Started, PP=Present Position

Monitor and control the project. It is important that project progress is consistently checked against the project objective. Monitoring and controlling aims at identifying and eliminating deviations from the project plan, scope, budget, duration, and quality. It is about taking corrective action when the need arises to reduce planned and actual gaps so that project goals are met. Measuring, recording, and collating, analyzing relevant, credible, and timely data that has a bearing on the project does this. The status column can be used to check off any aspect of the project that has been accomplished.

Close the Project. When the product or project deliverable is ready, it is handed over to the project owner. On the flip side, you would also terminate the project even if it fails. At this point, you would finalize all project activities. This means to end the project. You can do this by collecting all project records; determine if the project is successful or failed orderly terminate the project, project team, and project activities. Gather project materials, information, and experiences for lessons learned, produce a closure report and archive them for future use.

Project Failure and Lessons learned

Failure is not an option for project owners and for project managers. It is said that if you fail to plan, you have planned to fail. There is much more to project failure than meets the eye. However, we can glean a whole lot from records of lessons learned. It should, therefore, be paramount in our minds' prior experiences, and records can help us navigate better present project endeavors. Failure of future projects could be premised on when the project manager and the project team members ignore lessons from prior projects or are convinced that there are no lessons to be learned.

The point is situations that would require similar applications or maybe considerations from prior experiences should be well documented and archived for future reference. You cannot say you have fully closed a project if there is a lack in this area. A project that fails certainly has many possible reasons for the failure.

> **Project myopia** is a serious factor for project failure. However, meeting stakeholder and project expectations drive success.

Whether the sole or numerous reasons for failure were addressed by lessons learned in a previous project or not, it is apparent that the failure of a project could border on the fact that

- The project team was not aware of a problem, simply disregarded, or did not adequately utilize the lessons learned from such projects.
- There is project myopia or shortsightedness; that is project team members perceive the project as successful, while stakeholders and the external community think otherwise.
- The project scope, time, and cost were excessively surpassed, while quality was below expectations, resulting in non-financial viability of the deliverable.
- The project lacked the ability to implement project management methodologies that resulted in the delivery of lesser or no value product.

Nevertheless, success factors include when project deliverables meet scope, cost, and time expectations (including formal and approved changes), do not have shortcomings, and all stakeholders are satisfied and receive intended benefits.

Table 5.2
Do Projects Really Fail?

> According to the 1995 Standish Group Chaos Report, 90% of projects do not meet time, cost, and quality targets. On the average, only 9% of projects performed by large companies, 16.2% of software projects, and 28% of small company projects were considered successful. That is, they were completed on time, within budget, and delivered measurable business and stakeholder benefits. The reasons for such failures among others include lack of or inadequacy in user involvement, support form executive management, clear statement of requirements, proper planning, realistic expectations, competent staff, and hard working and focused staff.

Large and Complex Projects

By now, it is certainly clear which projects can be categorized as small or large, simple or complex. Some practitioners consider projects that are less than one million dollars in value as small. Others consider projects of more than one hundred thousand dollars in value as large. Categorization ought to be relative and dependent on the industry, as well as the objective of the project.

The rule of thumb is as discussed earlier and based on the fewness of skill areas, broadness of scope, and complexity of deliverable. Nevertheless, managing large and complex projects in terms of abiding by the principles of project management is by no means different from managing small and simple. It is the hugeness and involvement of more resources that separate the simple from the complex and the small from the large. Naturally, the employment, involvement, and deployment of more resources would introduce levels and hierarchies of complications.

Within certain limits, we can successfully utilize our knowledge of project management phases and process groups to manage small and simple projects. Beyond this, we will certainly require detailed knowledge of project management processes, more complex and automated tools, as well as the nine areas of the project management body of knowledge. This kind of knowledge provides you with information and understanding that you need to manage large and complex projects and to solve problems effectively.

It is also imperative for your general management knowledge and skill to be employed here, given that certain leadership and management attributes will overlap with the skill you require to manage projects effectively. Nevertheless, there are more activities within the process groups than we have discussed previously and definitely more from the nine knowledge areas, which in summary include:

1. Project integration management—ensures the capability to use various processes and methodologies to integrate and coordinate aspects of a project, so they can all work together rather than interfere with the common goal.

2. Project scope management—ensures the capability to consider and effectively manage a project's range, the extent of work, and the amount of resources required to complete a deliverable.
3. Project time management—ensures the capability to estimate, schedule, and manage processes and methodologies effectively for the timely completion of a project.
4. Project cost management—ensures the capability to utilize tools and techniques effectively to estimate the cost of resources in a way to complete a project within budget.
5. Project quality management—ensures the capability to use various project management processes and methodologies to maintain full compliance with quality standards, policies, or procedures of an organization during project performance.
6. Project human resource management—ensures the capability to motivate and control human resources involved in a project so that their effectiveness is increased during project performance while using the various project management processes and methodologies.
7. Project communication management—ensures the capability to utilize various processes and methodologies in establishing an effective system of information and communication flow among project team members and stakeholders, considering the relative complexity of a project.
8. Project risk management—ensures the capability to utilize various processes and methodologies proactively to identify, quantify, analyze, and respond to projects risks and uncertainties.
9. Project procurement management—ensures the capability to utilize various processes and methodologies to purchase and acquire goods and services for a project.

These tools and knowledge areas are discussed in more detail in Divisions 3 and 4 of this book. However, would it be apt to ask at this point if project management is still a cakewalk? Maybe it is or maybe it is not. But it is certainly one cake the recipe has been decomposed for easy learning. Nevertheless, any project will fail if its managers lack adequate knowledge and understanding of what needs to be done to manage a project successfully whether it is small or large, simple or complex.

> Detailed knowledge of the **nine knowledge areas** of project management provide information and understanding needed to manage large and complex projects and to solve project problems effectively

Test Your PM Knowledge

21. The difference between a small and large project does not include
    a. Management of skills areas
    b. Fewness of skill areas
    c. Broadness of scope
    d. Complexity of deliverable

22. Simple projects can equate to _____
    a. Small projects
    b. Assignments
    c. Complex project
    d. Duration projects

23. True or false: A project champion is the enthusiast for a project who convinces people or stakeholders to get them to see the importance of a project.
    a. True
    b. False

24. A project charter is one of the deliverables of a project's initiation phase. What is the output of the initiation phase?
    a. A projects organizational processes
    b. Statement of work
    c. A project deliverable
    d. A project charter

25. Failure is not an option for project owners nevertheless projects fail. Which is the most feasible reason why they fail?
    a. Project myopia
    b. Excessive surpassing of project cost, time, or scope
    c. Ignorance of project problem
    d. All of the above
    e. None of the above

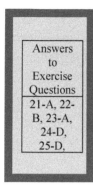

Answers to Exercise Questions
21-A, 22-B, 23-A, 24-D, 25-D,

Division 3

Project Management Stonecutters and Implements

Chapter 6

Project Tools and Techniques

- *Gantt Chart*
- *Flow Chart*
- *WBS, Decomposition*

Project Management Tools and Techniques

There are a huge number of tools and techniques for managing and facilitating each project management phase, process, and knowledge area.

Project Management Tools and Techniques

Acquisition	Cost Aggregation	Issue Logs Leads and Lags
Advertising	Cost Change Control	Lessons Learned Process
Alternative Identification	Cost of Quality	Make-Or-Buy Analysis
Alternative Analysis	Cost-Benefit Analysis	Management Teams
Analogous Estimating	Critical Path Method (CPM)	Negotiation—(Contract, HR, General)
Arrow Diagram Method (ADM)	Data Gathering and Representation	Networking
Benchmarking	Decomposition	Observation and Conversation
Bidder Conferences	Defect Repair Review	Organization Charts
Bottom-up Estimating	Dependency Determination	Organizational Theory
Parametric Estimating	Design of Experiments	Parametric Estimating
Cause and Effect Diagram	Diagramming Techniques	Pareto Chart
Change Control system—(Contract, Scope, others)	Documentation Reviews	Performance Appraisals
Checklist Analysis	Earned Value Technique	Performance Information Gathering/Compilation
Co-location	Expert Judgment	PERT
Communication Methods	Flowcharts	Performance Measurement Analysis
Communication Requirements Analysis	Forecasting	Planning Meetings and Analysis
Communications Skills	Funding Limit Reconciliation	Positions Description
Communications Technology	Gantt Chart	Pre-assignment
Configuration Management System	General Management Skills	Precedence Diagramming Method (PDM)
Conflict Resolution	Ground Rules	Probability and Impact Assessment/Matrix
Contingent Risk Response	Histogram	Process Analysis
Contract Types	Independent Estimates	
Control Charts	Information Gathering and Distribution	
	Information Presentation Techniques	
	Inspection	

Procurement Audits	Records Management	Statistical Sampling
Product Analysis	Replanning	Status Review
Project Management Information Systems	Reserve Analysis	Team-Building
	Resource Cost Rates Determination	Templates, Forms, and Standards
Project Management Methodology	Resource Leveling	Threat and Opportunity Strategies
Project Management Software	Risk Assessment	
	Risk Audits	Three Point Estimates
Project Performance Reviews	Rolling Wave Plan	Time and Cost Reporting
	Run Chart	Training
Project Selection Methods	Scatter Diagram	Trend Analysis
Proposal Evaluation Techniques	Schedule Network Templates	Variance Analysis
		Variance Management
Published Estimating Data	Screening System	Vendor Bid Analysis
Quality Audits	Sellers and Qualified Sellers List	Virtual Teams
Quality Control Tools		WBS Dictionary
Quality Planning Tools	Sellers Rating System	Weighting System
Quantities Risk Modeling	Stakeholder Analysis	Work Breakdown Structure (WBS)
Recognition and Rewards	Standard forms	

Figure 6.1 – Project Management Tools and Techniques

The advantage of using these tools and techniques is that, like implements, they ease the work of project management. Most of the tools and techniques are listed alphabetically in Figure 6.1. However, we will discuss only a few key ones.

The Gantt Chart

The Gantt chart, developed by Henry Gantt in 1910, is the oldest, simplest, most popular, and, if not, the best way of visually representing a project plan. The Gantt chart is a bar chart that shows the relationship of project activities over the project duration. In other words, it shows the project schedule. The chart is a matrix of columns and tables that vertically lists project activities on the far left column. It horizontally illustrates a time scale and horizontal bars—one for each activity that indicates their start and finish times in proportion to the length of each bar.

Considerable improvements have gone into the Gantt chart. While it shows the project schedule, some limitedly show dependency, that is, the reliance between the finish of one activity and start of another. Percent-complete shadings of the bars can also be used to show how much work within an activity has been completed. Generally, they are common in representing the phases, tasks, and activities of a project WBS—work breakdown structure. Note that scheduling your projects using Gantt charts should follow your activity definition and listing, and not the other way around.

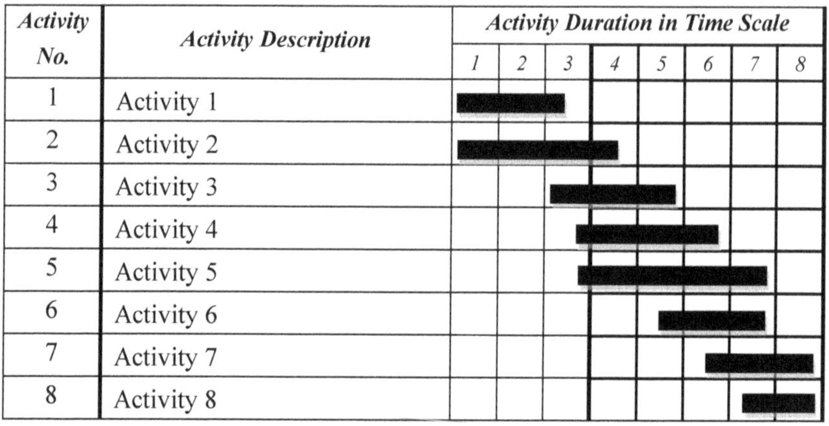

Activity No.	Activity Description	Activity Duration in Time Scale							
		1	2	3	4	5	6	7	8
1	Activity 1	■							
2	Activity 2		■	■					
3	Activity 3				■				
4	Activity 4					■	■		
5	Activity 5				■	■			
6	Activity 6						■		
7	Activity 7							■	
8	Activity 8								■

Figure 6.2 – The Gantt

Uses, Benefits, and Limitations of Gantt Charts

- They are used for project scheduling and project time management.
- They can be easily drawn, generated or developed using Excel spreadsheets, and they have low training demands.
- They cannot easily or clearly show task and activity dependencies in their plain form.
- They are most suited for managing small and simple projects.
- In situations where more information needs to be communicated, Gantt charts communicate relatively little.

Flowcharts

Considering the responsibility of project management to be able to visualize every aspect and process(es) of doing a project, flow charts are helpful. Flowcharts are schematics originally used in quality control to represent processes or algorithms. It is essential to employ flowcharts to help visualize your project's processes and content or its flaws better. They could also help in detailing project's scope better, thereby, providing you a schematic model of your project processes.

Within the realm of process models, we could consider two specific approaches in developing flowcharts that are suitable for managing projects. Although they may remain incomplete representations of processes, they will suffice for accomplishing our project tasks. They are the *workflow* type, which depicts processes, sometimes in graphical forms and as a set of workflows composed of activities that bear some relationship with themselves. The other is the *data or information flow* view, which is considered a traditional type of business process that involves data processing entities that have been used to understand and automate the flow of data. These are also represented in graphical forms as data flow diagrams (see Figure 6.2).

It is considered that activities that are not interrelated are not part of the process. Given that the structures of these processes are not real, they remain mental abstractions that enable understanding of the scope and boundaries of project functions. An example would be the set of activities that involve the design of a new product or the release of the newly developed software. What makes flowchart diagrams peculiar is that they manifest the ability to telescope huge systems and processes regardless of their size. They are able to represent different functions of time, human task, machine activity, and even people who are directly involved in processing activities that matter the most.

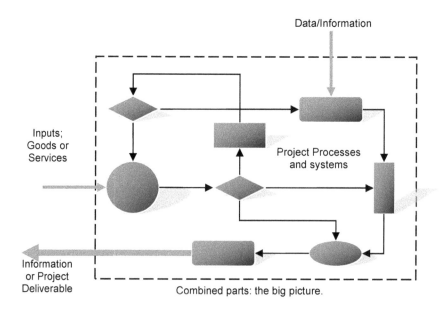

Figure 6.3 – Flowchart. Shows data and workflow, process input, output, parts, and relationships

Uses, Benefits, and Limitations of Flowcharts

Among many others, another significant benefit of using flowcharts in the view of experts and theorists is that they take care of the problem of *spatial blindness*—a source of considerable misunderstanding and conflict, especially when other parts of a project system are, for the most part, invincible mainly during analysis. So much so, when we consider that the whole is a function of its characteristic parts. Therefore, it will be inadequate to attempt to analyze or visualize a part of project system in isolation to the whole system.

In spite of the perceived limitations of the flowchart, it has proven to be very useful in the analysis of very complex project processes. For example, it is seeing the big picture where current events in a project's life can be related to its seemingly invincible history, as well as, seeing in contexts of project complexity and the responsibility of its management.

WBS, Decomposition

It might not be completely appropriate to say that project management can rely on any one tool or technique. However, the WBS—work breakdown structure— is a fundamental tool that focuses on the deliverable and helps us hierarchically decompose project work into more manageable units. We have discussed the systematic nature of project management and how it comprises of parts and subparts. The same goes for the holistic nature of projects. If we must succeed in executing our projects, then it is necessary that we reduce the work involved into parts or deliverables, sub-deliverables, and even smaller deliverables.

A projects' scope is very important to the successful execution of that project. The WBS organizes and defines the total scope of a project, identifies major issues, breaks them down into smaller deliverables and measurable units. This breakdown structure represents the deliverables in increasing descending levels and detailed definitions of phases, tasks, sub-tasks, and work packages.

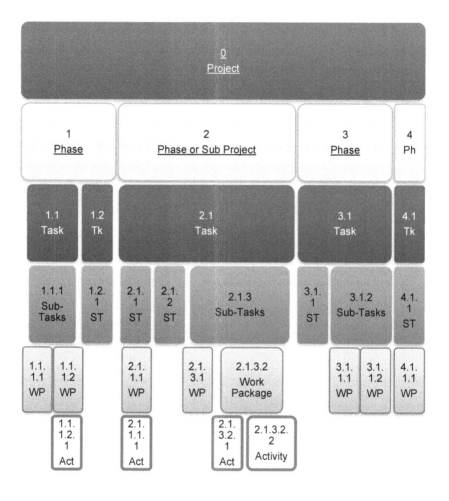

Figure 6.4a – WBS and Project Decomposition

During the development of the WBS, the different levels of work decomposition help us identify and define how to schedule, estimate the cost of, monitor, and control work packages. This means to identify and define; what tasks will be accomplished, who will do them, how long it will take, materials and supplies that are required, and how much each task will cost.

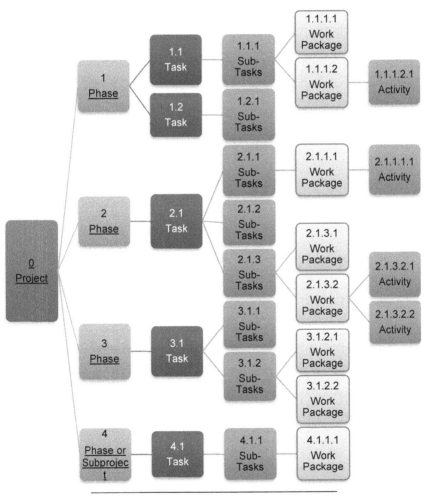

Figure 6.4b– WBS and Project Decomposition

Generally, the descending levels of work definition and decomposition have been structured around projects and programs as the first level. Subsequent levels are the project phases, tasks, sub-tasks, and work packages as the lowest levels. In specific instances, work packages have been further broken down during processes of activity definition (discussed further under project time management). The military-style numbering—that is 1, 1.1, 1.1.1, and so on— have been adopted for easy identification of these levels.

Uses, Benefits, and Limitations of WBS

It should be noted that project work not included in the WBS is not within the scope of the project given that it serves as an input into the scope change control process (discussed in detail under Project scope management). Other WBS uses include

- Identification of major project issues that provide the foundation for project planning and execution, monitoring and controlling.
- Breakdown of major tasks into smaller deliverables and subdivides them into measurable components for the purposes of accurately estimating activities, costing, scheduling, and assignment of work responsibility.
- Definition of each work package for the development of a shared understanding of the project scope, an organizational breakdown structure (OBS), a cost breakdown structure (CBS), and the iterative review of patterns and inconsistencies.

Note: *CPM, ADM, PERT, Product Analysis, Expert Judgment, Earned Value Management, and other Tools and Techniques will be presented in BOOK 2.*

Test Your PM Knowledge

26. The Gantt Chart was developed in 1910 by Henry Gantt to represent a project plan visually. Project managers simply use it to
    a. Schedule projects and manage time
    b. Show task dependencies
    c. Communicate a lot of project information
    d. Manage large projects.

27. Flowcharts are helpful schematics originally used in quality control. Project managers have adopted them to help with
    a. Visualizing every part of a project system
    b. Analyzing parts of a project system while seeing visualizing the whole
    c. Options (a) and (b) are correct
    d. Options (a) and (b) are incorrect

28. True or false: Flowchart diagrams do not have the peculiar ability to telescope huge systems and processes regardless of their size.
    a. True
    b. False

29. Gantt Charts are most suited for managing _____ projects and they vertically list project _____ on the far _____ column.
    a. Small, schedule, left
    b. Simple, schedule, left
    c. Small, activities, left
    d. Simple, activities, right

30. The WBS is one of project management's fundamental tools that focuses on the _____ and is used to _____ project work into more _____ units.
    a. Deliverable, decompose, manageable
    b. Organization, plan, systematic
    c. Work packages, assemble, organized
    d. Tasks, arrange, deliverable

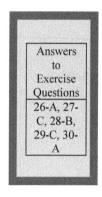

Answers to Exercise Questions
26-A, 27-C, 28-B, 29-C, 30-A

Chapter 7

Project Management Technology

- *Project Logbook*
- *Project Management Software*

Over time, the task of managing projects has become a breeze with the use of available but advanced technology. It is not as effortless as it sounds; however, relevant technology has made managing especially complex projects a great deal easier. This is not without acquiring the skill to use them efficiently and effectively.

Nevertheless, there is a danger to over-dependence on these technologies, that is if you lack requisite project management knowledge and skills. The technologies will not think or make decisions for you, especially, if you know that the MO—modus operandi—of these technologies is *garbage in garbage out*. However, they will help you keep records and process information faster and with relative levels of accuracy.

Project Logbook

As traditional as this may sound, the project logbook is a very pragmatic way and effective way to log all project information. This is because it serves as your project register. It provides a practical way to organize for easy access, all your project data, information, and documentation. The logbook should include project charter, scope statement, WBS, schedules, status reports, risk and response plans, baselines, change requests and others.

The size of your project logbook should be determined based on what is convenient and what works for you. However, if it is necessary that your carry it around, then choose a logbook that is not too small to get lost, but not too bulky to carry around. We can also not be too traditional with the format of the logbook. There are PDA's—personal digital assistances—and electronic books that can carry the same if not more information. In addition, you can always synchronize or exchange project information between your desktop computer and your PDA or mobile devices.

Although your PDA may not have adequate capacity for project management software, it could carry around certain important files you need to have handy. For example, such files can be in PDF, Word, and Excel formats. Check your PDA functions and electronic book operating systems for the kind of software applications they are compatible with.

Project Management Software

Up until now in our discussions, the different tools and techniques used in managing projects have been and could be created manually. This is okay for simpler projects, but as a project tends towards complexity, we need the advantage of advanced technology for more efficient coordination. Project management software delineates a range of computer software that can be used for time and budget management, scheduling and resource allocation, quality and procurement management, communication and documentation, and others.

There are also various platforms for using and implementing project management software. They include personal, desktop, web-based, integrated, and collaborative software application platforms.

- Personal Software—is relatively simpler single user applications for managing simple projects. Spreadsheets and calendar software fall under this category.
- Desktop Applications—can be used for personal and office project management. They can be used to process and store project information that are exclusive to the desktop user(s) or shared over a network with others. Offices can also integrate other office

programs with project management software. Such human resource information is available and accessible to a project manager.

- Web-based Applications—have the advantage of accessibility and multi-use by anyone granted access over the internet or extranet with a web browser. Web-based applications can also provide a basis for multi-user collaboration in which users can add information or modify sections of project information.

Types and Categories of Project Management Software

Available project management software is used for various project management functions. There is none that can do all. However, notable applications are categorized as open-source desktop and web-based and propriety desktop and web-based. Open-source software is a free—usually publicly and collaboratively developed—computer software that permits the user under a copyright license to modify, improve, or redistribute the software in the modified or unmodified form. On the contrary, propriety software restricts use or private modification, improvement, or redistribution. The proprietors of the software are at liberty to distribute such software free or at a fee.

Open-source project management (desktop and web-based) software include

- GanttProject—is a desktop application developed by the GanttProject Team (www.ganttproject.biz) with Gantt chart, task hierarchy, task dependency, and resource load chart features. They can also generate PERT charts; produce PDF and HTML reports, exchange data with spreadsheet applications, such as Excel, and import and export to and from MS Project.

- OpenProj—is a relatively new and popular desktop application that was initially released in 2006. Developed by Projity (www.projity.com), its features include earned value costing, Gantt and PERT chart, Task usage reports, resource breakdown structure (RBS), and work breakdown structure (WBS).
- Project.net—is a web-based application developed by Project.net (www.project.net) for an enterprise-scale project and portfolio management. Three of its key features are collaboration (which helps people involved in a project achieve their common objective), resource management, and issue tracking (which helps with the management and maintenance of a list of project and people issues).
- dotProject—is a web-based multi-user, multi-language application. It is developed by dotproject (www.dotproject.net); its features include task management and contacts, to-do list, resource management, Gantt chart, discussion forums, reporting, calendar, and historical documentation.

Although there are more open source project management software applications, there is much more **propriety project management** (desktop and web-based) applications. Some include

- Microsoft Project—is a desktop application developed and sold by Microsoft (http://office.microsoft.com). Its features include assistance in developing plans, resource assignment to tasks, scheduling of tasks, tracking of project progress, managing budgets, and workload analysis.
- Artemis Views—is a desktop enterprise-wide (documents and collaboration) management application developed by Artemis International Solutions Corporation

(www.artemispm.com/product/2). The software is claimed to have more than 500,000 users for advanced project planning and management reporting among others. They can also be used for time reporting, project analytics, earned value, and management.

- Project Engine—is a combined enterprise project planning and issue tracking desktop application developed in Sweden by Innovative Tool Solutions (www.projectengine.se). Its features include a task tree, distributed to-do list, workflow engine, instant messaging, software version control client, and ability to access tasks from any location.
- Daptiv—is a web-based collaborative project management application developed by Daptiv, Inc. (www.daptiv.com) for resource, task, and portfolio management, issue and time tracking, and document management.
- ProjectPartner—is a web-based portfolio and project management software developed by Timedisciple (www.timedisciple.com). Its features include timesheet recording of project activities, tasks, and roles, tasks, phases, resources, and budget management.
- Teamwork—is a web-based application for managing work and communication used in an integrated and collaborative web environment. Developed in Italy by Open Lab (www.twproject.com), its features include time and issue tracking, ability to coordinate and manage hundreds of projects at once, ability to integrate collaborative social tools and email clients, and ability link documents without moving files.

There is more project management software than we have mentioned here. You should do more searches for the ones that possess functionalities that will satisfy your project organization.

Test Your PM Knowledge

31. Availability of advanced technology has made managing projects a lot _____
    a. More difficult
    b. Less difficult
    c. More complex
    d. Less efficient

32. A project logbook is a pragmatic and effective way to log all project _____
    a. Cash flow
    b. Risks
    c. Information
    d. schedules

33. There are various platforms for using and implementing project management software. Which of the following is not one of the platforms?
    a. Personal Software
    b. Desktop Applications
    c. Web-based Applications
    d. Electronic logs

34. There are two main categories of software for performing project management functions. They are
    a. Open source project management software
    b. Only option (a) is correct
    c. Propriety project management software`
    d. Options (a) and (c) are correct

35. Is it true that propriety software's restrict use or private modification and improvement?
    a. Yes
    b. No
    c. Don't know
    d. Not sure

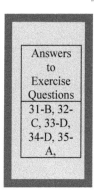

Answers to Exercise Questions
31-B, 32-C, 33-D, 34-D, 35-A,

Chapter 8

Project Management Influences

- *Project Organization*
- *Project Management Triple Constraints*
- *Project Processes and Knowledge Interaction*

It should not go without recognizing the behavioral tendency of certain things—internal and external to a project—and the power they exert over our projects. In the beginning chapters of this book, we considered different factors and environments we need to be aware of in managing projects. However, projects are usually part of an organization, which means that we amalgamate *information*, *people*, and *resources* to reach project goals. The operation, dynamism, and interplay of these three components greatly affect the outcome of our projects.

> The **Project Organization** is a structure or instrument we create to give project stakeholders and team members the power needed to shape and control project work and people's behavior to produce, deliver, and accept project deliverables.

Project Organization

Organizations are complex systems of people that utilize a variety of resources to accomplish a predetermined purpose. As a body and consolidation of persons, there are roles, responsibilities, and functions performed by each individual member of the organization through administrative or executive structures. There are also leaders, managers, policies, and procedures that inspire, direct, control, and guide how we perform organizational functions. Members of organizational systems bring their skills, abilities, creativity, experience, and possibly their needs to utilize information and resources—equipment, technology, capital, and time for the production of goods and services.

Types of organizations include public and private institutions, such as government agencies, corporations, international organizations, and professional associations. Directly or indirectly, our projects are connected to these kinds of organizations. If we consider that they are larger than our projects, they certainly exert some level of influence on our projects. Nevertheless, there is much in common between projects and organizations, in terms of both having objectives, structures, procedures, and managers

We must organize our projects around organizational management procedures and structures that are well-founded. This means that organizational structures should be adequate and cooperative enough to support our project actions. To this end, there are mainly three different kinds of organizational structures: the functional, the matrix or hybrid, and the projectized.

Figure 8.1 – Functional Organizational Structure

Figure 8.1, shows the classical functional organization. In this sort of structures members of the organization are grouped according to specialty and functions, such as human resources, finance, production, and others. Functional structures grant projects access to functional skills and management experience. Project systems and functional procedures are to some extent compatible, as project team members are committed to the objectives of the functional organization.

However, the limitation of this kind of structure is that functional managers coordinate projects, given that their priorities lie with managing their day-to-day operations. In addition, key project personnel are not available or are part-time, and project scope is limited to the boundaries of the functional managers.

In figure 8.2, team members are collocated in the projectized structure or organization, as organizational resources are deployed for project work. Project coordination is a full-time function and responsibility of managers—who have independence, authority, and team members—who are committed to project success.

> **Groupthink** is the tendency of members of a group, committee, or profession to manifest the same kind of thoughts such that opinions and feelings of group members conform to the dominant thought therefore overriding a realistic appraisal of a situation.

Contrary to functional organizations, roles can be duplicated between projects. Projects system and organizational procedures can be incompatible, and the end of a project could mean loss of a skill or experience gained during a project. Note, however, the danger of *groupthink* that such a homogenous group could suffer, that is, the danger of elevating its views over the needs of project stakeholders.

Some practitioners consider the matrix—a hybrid of the functional and projectized—structure and organization the best form of project organization. This is because it combines the best of both the functional and the projectized structures. The application of matrix organization can vary from the weak to the balanced, to the strong, and to the composite—which applies the matrix structure at various levels of the organization. These variables define the level of authority of a project manager in the organization.

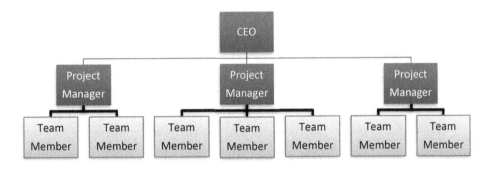

Figure 8.2 – Projectized Organizational Structure

The matrix is weak when the project manager is regarded merely as the project coordinator or expediter and reports to a functional manager. There is relatively no difference between the weak matrix and the functional. In the same vein, the strong matrix has the many characteristics of the projectized organization. Project managers are full time, have full-time staff, and act with more independence and authority.

The balanced matrix takes into consideration the need of a project manager but reports to a functional manager who has full authority over the project and its funding. The use of the matrix structure means that projects have access to resources and expertise of people from other departments and that project systems and organizational procedures are compatible. However, the major problem with the matrix structure is the confusion and conflict that would arise from project team members reporting to both a functional and a project manager.

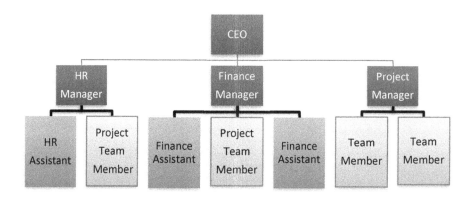

Figure 8.3 – Matrix Organizational Structure

Nevertheless, if you were to choose and decide the type of project organization most suitable for your project, top most in your consideration should include the type of organization that would

- Enable you to accomplish your project goals successfully.
- Aid you to utilize new and available technologies, machinery, and equipment.
- Grant you (possibly unhindered) access to functional and project skills, experience, and abilities.
- Give you the independence and relevant authority to manage full time functional and project staff.

- Meet and satisfy the expectations of your stakeholders.
- Support your effort to finish your project and deliver the product within the project scope, cost, and expected time.

A project's scope, time, and cost are three constraints that every project has to deal with. Known as the triple constraint—traditionally called the project management triangles—they all have significant change effects on each other. The way they interact is discussed more in the next section.

Projects Triple Constraints

The three sides of an equilateral triangle are used to represent the triple constraints of a project. Quality is considered an implied fourth constraint because of the impact it has on the traditional three. Such impact is a function of the thinking that project quality is considered in making decisions throughout every phase, process, and life of the project. At the base of the triangle is scope, which defines the totality and work boundaries of a project's product or services or both. However, the relationship of the scope to both cost and time is that the latter is computed based on the project scope.

Figure 8.4 – The Triple Constraints

Nevertheless, the idea of representing them equally on an equilateral triangle is to establish a balance of importance of all three constraints, especially at the beginning of a project. As the project progresses, changes may be noticed as one, two, or even all three constraints are surpassed or decreased. The nature of change is represented in relationship triangles illustrated in figures 8.4 to 8.7. Changes in scope—that is, what must be done to produce a project deliverable—could be magnified or shrunken. This will have corresponding effects on the project completion time and budget.

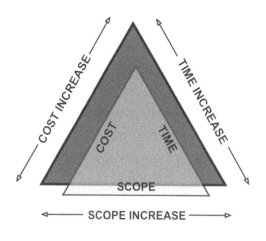

Figure 8.5 – Increase in all Three Constraints

Generally, a reduced cost constraint could mean an equally reduced scope and reduced time, or on the contrary, an increased scope means increased cost and could imply increased completion time. See figure 8.5. However, a reduced completion time while scope is the same could imply an increased cost. In the same manner, a reduced budget while scope is the same could mean that the project will take a much longer time to complete. See Figure 8.6.

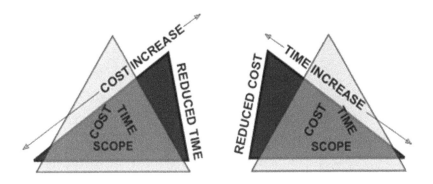

Figure 8.6 – Increased cost with an unchanged scope can lead to reduced time. The reduced cost can mean increased time with unchanged scope.

Project Processes and Knowledge Activities Interaction

There are varieties of activities that are performed during project management. They occur while project process groups interact with specific knowledge areas. The activities include identification, analysis, and design of objectives; planning, estimating, and allocation of resources; acquiring human resources, assigning tasks, directing activities; executing, controlling and monitoring, quality control, communicating and closing. A matrix of the interaction of such activities is shown in figure 8.7.

Project Management Areas & Processes	Project Management Process Groups				
	Initiating	Planning	Executing	Monitoring & Controlling	Closing
Integration	●	●	●	●	●
Scope		●		●	
Time		●		●	
Cost		●		●	
Quality		●	●	●	
Human Resource		●	●	●	
Communications		●	●	●	
Risk		●		●	
Procurement		●	●	●	●

Figure 8.7 – Project Management Processes and Knowledge Interaction Matrix

Test Your PM Knowledge

36. True or false: Projects are directly or indirectly connected to some public or private intuitions, which exert some level of influence on the project.
    a. False
    b. True

37. In the functional organization structure of a project organization, members are grouped according to their _____
    a. Profession
    b. Education
    c. Specialty
    d. Authority

38. In the projectized organizational structure of an organization ___
    a. Members are collocated as resources are deployed for project work
    b. Project coordination is a full time responsibility of managers
    c. Managers have independence and authority over team members
    d. All of the above

39. What sort of group or committee will most probably not suffer from groupthink?
    a. A committee of teachers and traders
    b. A committee of medical doctors and pharmacists
    c. A committee of architects and structural engineers
    d. A committee of bankers and lenders

40. The best form of project organization is the one with a _____ organizational structure.
    a. Functional
    b. Executive
    c. Projectized
    d. Matrix

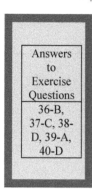

Answers to Exercise Questions
36-B, 37-C, 38-D, 39-A, 40-D

Bibliography

Baguley p. (1999). *Project management.* London, UK: Hodder Headline Plc.

Champoux, J. E. (2006). *Organizational behavior: Integrating individuals, groups, and organizations* (3rd ed.). Mason, OH: Thomson.

Hitt, M. A., Black, J.S., & Porter L. W. (2009). *Management* (2nd ed.). Upper Saddle River, New Jersey: Pearson Prentice Hall.

Kemp, S. (2004). *Project management demystified.* New York, NY: McGraw-Hill.

Odiaka, E. (2008). *Process and performance optimization strategies by labor intensive small- and medium-sized industries: Cross cultural cases from America and Nigeria.* (Doctoral dissertation, Walden University, 2008). UMI/Proquest No. 3297178

Oshry, B. (1996). *Seeing systems: Unlocking the mysteries of organizational life.* San Francisco: Berrett-Koehler Publishers.

PM College (2005). Professional development personal lesson notes [from PM professional coach e-learning program @ www.pmcollege.com]

Project Management Institute (2004). *A guide to the project management body of knowledge: PMBOK® guide–* 3rd ed. Pennsylvania, USA: Project Management Institute, Inc.

Project Management Institute (2008). *A guide to the project management body of knowledge: PMBOK® guide–*fourth edition. Pennsylvania, USA: Project Management Institute, Inc.

Rowe S. F. (2007). *Project management for small projects.* Vienna, VA: Management Concepts.

The Standish Group Report (1995). *Chaos.* Retrieved on 05/12/2008, from http://net.educause.edu/ir/library/pdf/NCP08083B.pdf

Whitten, J., & Bentley, L. (2007). *Systems analysis and design methods* (7th Ed.). New York: McGraw-Hill/Irwin.

Glossary

Included under permission from Project Management Institute *A guide to the project management body of knowledge (PMBOK ®Guide)* – Fourth Edition, Project management Institute, Inc., 2008. This glossary mostly includes a preponderance of terms that have and have not been mentioned in this book but are unique or nearly unique to the project management profession and body of knowledge. A few others are not unique but are used differently in general everyday use.

Activity. A component of work performed during the course of a project.

Activity Duration. The time in calendar units between the start and finish of a schedule activity. See also duration.

Activity Identifier. A short unique numeric or text identification assigned to each schedule activity to differentiate that project activity from other activities. Typically unique within any one project schedule network diagram.

Activity List [Output/Input]. A documented tabulation of schedule activities that shows the activity description, activity identifier, and a sufficiently detailed scope of work description so project team members understand what work is to be performed.

Actual Cost (AC). Total costs actually incurred and recorded in accomplishing work performed during a given time period for a schedule activity or work breakdown structure component. Actual cost can sometimes be direct labor hours alone, direct costs alone, or all costs including indirect costs. Also referred to as the actual cost of work performed (ACWP). See also earned value management and earned value technique.

Analogous Estimating [Technique]. An estimating technique that uses the values of parameters, such as scope, cost, budget, and duration or measures of scale such as size, weight, and complexity from a previous, similar activity as the basis for estimating the same parameter or measure for a future activity.

Approved Change Request [Output/Input]. A change request that has been processed through the integrated change control process and approved.

Baseline. An approved plan for a project, plus or minus approved changes. It is compared to actual performance to determine if performance is within acceptable variance thresholds. Generally refers to the current baseline, but may refer to the original or some other baseline. Usually used with a modifier (e.g., cost performance baseline, schedule baseline, performance measurement baseline, technical baseline).

Bottom-up Estimating [Technique]. A method of estimating a component of work. The work is decomposed into more detail. An estimate is prepared of what is needed to meet the requirements of each of the lower, more detailed pieces of work, and these estimates are then aggregated into a total quantity for the component of work. The accuracy of bottom-up estimating is driven by the size and complexity of the work identified at the lower levels.

Brainstorming [Technique]. A general data gathering and creativity technique that can be used to identify risks, ideas, or solutions to issues by using a group of team members or subject-matter experts.

Budget. The approved estimate for the project or any work breakdown structure component or any schedule activity. See also estimate.

Change Control. Identifying, documenting, approving or rejecting, and controlling changes to the project baselines.

Change Control Board (CCB). A formally constituted group of stakeholders responsible for reviewing, evaluating, approving, delaying, or rejecting changes to a project, with all decisions and recommendations being recorded.

Change Control System [Tool]. A collection of formal documented procedures that define how project deliverables and documentation will be controlled, changed, and approved. In most application areas, the change control system is a subset of the configuration management system.

Change Request. Requests to expand or reduce the project scope, modify policies, processes, plans, or procedures, modify costs or budgets, or revise schedules.

Charter. See project charter.

Close Project or Phase [Process]. The process of finalizing all activities across all of the Project Management Process Groups to formally complete the project or phase.

Closing Processes [Process Group]. Those processes performed to finalize all activities across all Project Management Process Groups to formally close the project or phase.

Communication Management Plan [Output/Input]. The document that describes: the communications needs and expectations for the project; how and in what format information will be communicated; when and where each communication will be made; and who is responsible for providing each type of communication. The communication management plan is contained in, or is a subsidiary plan of, the project management plan.

Constraint [Input]. The state, quality, or sense of being restricted to a given course of action or inaction. An applicable restriction or limitation, either internal or external to a project, which will affect the performance of the project or a process. For example, a schedule constraint is any limitation or restraint placed on the project schedule that affects when a schedule activity can be scheduled and is usually in the form of fixed imposed dates.

Contingency. See reserve.

Contingency Allowance. See reserve.

Contingency Reserve [Output/Input]. The amount of funds, budget, or time needed above the estimate to reduce the risk of overruns of project objectives to a level acceptable to the organization.

Contract [Output/Input]. A contract is a mutually binding agreement that obligates the seller to provide the specified product or service or result and obligates the buyer to pay for it.

Control. Comparing actual performance with planned performance, analyzing variances, assessing trends to effect process improvements, evaluating possible alternatives, and recommending appropriate corrective action as needed.

Control Chart [Tool]. A graphic display of process data over time and against established control limits, and that has a centerline that assists in detecting a trend of plotted values toward either control limit.

Control Limits. The area composed of three standard deviations on either side of the centerline, or mean, of a normal distribution of data plotted on a control chart that reflects the expected variation in the data. See also specification limits.

Control Schedule [Process]. The process of monitoring the status of the project to update project progress and managing changes to the schedule baseline.

Control Scope [Process]. The process of monitoring the status of the project and product scope and managing changes to the scope baseline.

Controlling. See control.

Corrective Action. Documented direction for executing the project work to bring expected future performance of the project work in line with the project management plan.

Cost Management Plan [Output/Input]. The document that sets out the format and establishes the activities and criteria for planning, structuring, and controlling the project costs. The cost management plan is contained in, or is a subsidiary plan of, the project management plan.

Cost of Quality (COQ) [Technique]. A method of determining the costs incurred to ensure quality. Prevention and appraisal costs (cost of conformance) include costs for quality planning, quality control (QC), and quality assurance to ensure compliance to requirements (i.e., training, QC systems, etc.). Failure costs (cost of nonconformance) include costs to rework products, components, or processes that are non-compliant, costs of warranty work and waste, and loss of reputation.

Cost Performance Baseline. A specific version of the time-phased budget used to compare actual expenditures to planned expenditures to determine if preventive or corrective action is needed to meet the project objectives.

Cost Performance Index (CPI). A measure of cost efficiency on a project. It is the ratio of earned value (EV) to actual costs (AC). CPI = EV divided by AC.

Cost-Plus-Fixed-Fee (CPFF) Contract. A type of cost-reimbursable contract where the buyer reimburses the seller for the seller's allowable costs (allowable costs are defined by the contract) plus a fixed amount of profit (fee).

Cost-Plus-Incentive-Fee (CPIF) Contract. A type of cost-reimbursable contract where the buyer reimburses the seller for the seller's allowable costs (allowable costs are defined by the contract), and the seller earns its profit if it meets defined performance criteria.

Cost-Reimbursable Contract. A type of contract involving payment to the seller for the seller's actual costs, plus a fee typically representing seller's profit. Cost-reimbursable contracts often include incentive clauses where, if the seller meets or exceeds selected project objectives, such as schedule targets or total cost, then the seller receives from the buyer an incentive or bonus payment.

Cost Variance (CV). A measure of cost performance on a project. It is the difference between earned value (EV) and actual cost (AC). CV = EV minus AC.

Crashing [Technique]. A specific type of project schedule compression technique performed by taking action to decrease the total project schedule duration after analyzing a number of alternatives to determine how to get the maximum schedule duration compression for the least additional cost. Typical approaches for crashing a schedule include reducing schedule activity durations and increasing the assignment of resources on schedule activities. See also fast tracking and schedule compression.

Create WBS (Work Breakdown Structure) [Process]. The process of subdividing project deliverables and project work into smaller, more manageable components.

Criteria. Standards, rules, or tests on which a judgment or decision can be based, or by which a product, service, result, or process can be evaluated.

Critical Activity. Any schedule activity on a critical path in a project schedule. Most commonly determined by using the critical path method. Although some activities are "critical," in the dictionary sense, without being on the critical path, this meaning is seldom used in the project context.

Critical Path. Generally, but not always, the sequence of schedule activities that determines the duration of the project. It is the longest path through the project. See also critical path methodology.

Critical Path Methodology (CPM) [Technique]. A schedule network analysis technique used to determine the amount of scheduling flexibility (the amount of float) on various logical network paths in the project schedule network, and to determine the minimum total project duration. Early start and finish dates are calculated by means of a forward pass, using a specified start date. Late start and finish dates are calculated by means of a backward pass, starting from a specified completion date, which sometimes is the project early finish date determined during the forward pass calculation. See also critical path.

Define Activities [Process]. The process of identifying the specific actions to be performed to produce the project deliverables.

Define Scope [Process]. The process of developing a detailed description of the project and product.

Deliverable [Output/Input]. Any unique and verifiable product, result, or capability to perform a service that must be produced to complete a process, phase, or project. Often used more narrowly in reference to an external deliverable, which is a deliverable that is subject to approval by the project sponsor or customer. See also product and result.

Delphi Technique [Technique]. An information gathering technique used as a way to reach a consensus of experts on a subject. Experts on the subject participate in this technique anonymously. A facilitator uses a questionnaire to solicit ideas about the important project points related to the subject. The responses are summarized and are then recirculated to the experts for further comment. Consensus may be reached in a few rounds of this process. The Delphi technique helps reduce bias in the data and keeps any one person from having undue influence on the outcome.

Dependency. See logical relationship.

Duration (DU or DUR). The total number of work periods (not including holidays or other nonworking periods) required to complete a schedule activity or work breakdown structure component. Usually expressed as workdays or workweeks. Sometimes incorrectly equated with elapsed time. Contrast with effort.

Earned Value (EV). The value of work performed expressed in terms of the approved budget assigned to that work for a schedule activity or work breakdown structure component. Also referred to as the budgeted cost of work performed (BCWP).

Earned Value Management (EVM). A management methodology for integrating scope, schedule, and resources, and for objectively measuring project performance and progress. Performance is measured by determining the budgeted cost of work performed (i.e., earned value) and comparing it to the actual cost of work performed (i.e., actual cost).

Earned Value Technique (EVT) [Technique]. A specific technique for measuring the performance of work and used to establish the performance measurement baseline (PMB).

Estimate [Output/Input]. A quantitative assessment of the likely amount or outcome. Usually applied to project costs, resources, effort, and durations and is usually preceded by a modifier (i.e., preliminary, conceptual, feasibility, order-of-magnitude, definitive). It should always include some indication of accuracy (e.g., ± x percent). See also budget and cost.

Estimate Activity Durations [Process]. The process of approximating the number of work periods needed to complete individual activities with estimated resources.

Estimate Activity Resources [Process]. The process of estimating the type and quantities of material, people, equipment or supplies required to perform each activity.

Execute. Directing, managing, performing, and accomplishing the project work, providing the deliverables, and providing work performance information.

Executing Processes [Process Group]. Those processes performed to complete the work defined in the project management plan to satisfy the project objectives.

Expert Judgment [Technique]. Judgment provided based upon expertise in an application area, knowledge area, discipline, industry, etc. as appropriate for the activity being performed. Such expertise may be provided by any group or person with specialized education, knowledge, skill, experience, or training.

Fast Tracking [Technique]. A specific project schedule compression technique that changes network logic to overlap phases that would normally be done in sequence, such as the design phase and construction phase, or to perform schedule activities in parallel. See also crashing and schedule compression.

Firm-Fixed-Price (FFP) Contract. A type of fixed price contract where the buyer pays the seller a set amount (as defined by the contract), regardless of the seller's costs.

Fixed-Price-Incentive-Fee (FPIF) Contract. A type of contract where the buyer pays the seller a set amount (as defined by the contract), and the seller can earn an additional amount if the seller meets defined performance criteria.

Float. Also called slack. See total fl oat and free fl oat.

Flowcharting [Technique]. The depiction in a diagram format of the inputs, process actions, and outputs of one or more processes within a system.

Functional Manager. Someone with management authority over an organizational unit within a functional organization. The manager of any group that actually makes a product or performs a service. Sometimes called a line manager.

Functional Organization. A hierarchical organization where each employee has one clear superior, and staff are grouped by areas of specialization and managed by a person with expertise in that area.

Gantt Chart. [Tool] A graphic display of schedule-related information. In the typical bar chart, schedule activities or work breakdown structure components are listed down the left side of the chart, dates are shown across the top, and activity durations are shown as date-placed horizontal bars.

Historical Information. Documents and data on prior projects including project fi les, records, correspondence, closed contracts, and closed projects.

Identify Stakeholders [Process]. The process of identifying all people or organizations impacted by the project, and documenting relevant information regarding their interests, involvement, and impact on project success.

Initiating Processes [Process Group]. Those processes performed to define a new project or a new phase of an existing project by obtaining authorization to start the project or phase.

Input [Process Input]. Any item, whether internal or external to the project that is required by a process before that process proceeds. May be an output from a predecessor process.

Lag [Technique]. A modification of a logical relationship that directs a delay in the successor activity. For example, in a finish-to-start dependency with a ten-day lag, the successor activity cannot start until ten days after the predecessor activity has finished. See also lead.

Lessons Learned [Output/Input]. The learning gained from the process of performing the project. Lessons learned may be identified at any point. Also considered a project record, to be included in the lessons learned knowledge base.

Life Cycle. See project life cycle.

Log. A document used to record and describe or denote selected items identified during execution of a process or activity. Usually used with a modifier, such as issue, quality control, action, or defect.

Manage Project Team [Process]. The process of tracking team member performance, providing feedback, resolving issues, and managing changes to optimize project performance.

Manage Stakeholder Expectations [Process]. The process of communicating and working with stakeholders to meet their needs and addressing issues as they occur.

Matrix Organization. Any organizational structure in which the project manager shares responsibility with the functional managers for assigning priorities and for directing the work of persons assigned to the project.

Milestone. A significant point or event in the project.

Monitor. Collect project performance data with respect to a plan, produce performance measures, and report and disseminate performance information.

Monitor and Control Project Work [Process]. The process of tracking, reviewing, and regulating the progress to meet the performance objectives defined in the project management plan.

Monitoring and Controlling Processes [Process Group]. Those processes required to track, review, and regulate the progress and performance of the project, identify any areas in which changes to the plan are required, and initiate the corresponding changes.

Network. See project schedule network diagram.

Network Path. Any continuous series of schedule activities connected with logical relationships in a project schedule network diagram.

Objective. Something toward which work is to be directed, a strategic position to be attained, or a purpose to be achieved, a result to be obtained, a product to be produced, or a service to be performed.

Organizational Breakdown Structure (OBS) [Tool]. A hierarchically organized depiction organization arranged so as to relate the work packages to the performing organizational units. of the project Organizational Process Assets [Output/Input]. Any or all process related assets, from any or all of the organizations involved in the project that are or can be used to influence the project's success. These process assets include formal and informal plans, policies, procedures, and guidelines. The process assets also include the organizations' knowledge bases such as lessons learned and historical information.

Output [Process Output]. A product, result, or service generated by a process. May be an input to a successor process.

Parametric Estimating [Technique]. An estimating technique that uses a statistical relationship between historical data and other variables (e.g., square footage in construction, lines of code in software development) to calculate an estimate for activity parameters, such as scope, cost, budget, and duration. An example for the cost parameter is multiplying the planned quantity of work to be performed by the historical cost per unit to obtain the estimated cost.

Pareto Chart [Tool]. A histogram, ordered by frequency of occurrence, that shows how many results were generated by each identified cause.

Performance Measurement Baseline. An approved integrated scope-schedule-cost plan for the project work against which project execution is compared to measure and manage performance. Technical and quality parameters may also be included.

Phase. See project phase.

Planned Value (PV). The authorized budget assigned to the scheduled work to be accomplished for a schedule activity or work breakdown structure component. Also referred to as the budgeted cost of work scheduled (BCWS).

Planning Processes [Process Group]. Those processes performed to establish the total scope of the effort, define and refine the objectives, and develop the course of action required to attain those objectives.

Portfolio. A collection of projects or programs and other work that are grouped together to facilitate effective management of that work to meet strategic business objectives. The projects or programs of the portfolio may not necessarily be interdependent or directly related.

Portfolio Management [Technique]. The centralized management of one or more portfolios, which includes identifying, prioritizing, authorizing, managing, and controlling projects, programs, and other related work, to achieve specific strategic business objectives.

Precedence Diagramming Method (PDM) [Technique]. A schedule network diagramming technique in which schedule activities are represented by boxes (or nodes). Schedule activities are graphically linked by one or more logical relationships to show the sequence in which the activities are to be performed.

Predecessor Activity. The schedule activity that determines when the logical successor activity can begin or end.

Preventive Action. A documented direction to perform an activity that can reduce the probability of negative consequences associated with project risks.

Probability and Impact Matrix [Tool]. A common way to determine whether a risk is considered low, moderate, or high by combining the two dimensions of a risk: its probability of occurrence and its impact on objectives if it occurs.

Product. An artifact that is produced, is quantifiable, and can be either an end item in itself or a component item. Additional words for products are material and goods. Contrast with result. See also deliverable.

Product Life Cycle. A collection of generally sequential, non-overlapping product phases whose name and number are determined by the manufacturing and control needs of the organization. The last product life cycle phase for a product is generally the product's retirement. Generally, a project life cycle is contained within one or more product life cycles.

Product Scope. The features and functions that characterize a product, service, or result.

Product Scope Description. The documented narrative description of the product scope.

Program. A group of related projects managed in a coordinated way to obtain benefits and control not available from managing them individually. Programs may include elements of related work outside of the scope of the discrete projects in the program.

Program Evaluation and Review Technique (PERT). A technique for estimating that applies a weighted average of optimistic, pessimistic, and most likely estimates when there is uncertainty with the individual activity estimates.

Program Management. The centralized coordinated management of a program to achieve the program's strategic objectives and benefits.

Project. A temporary endeavor undertaken to create a unique product, service, or result.

Project Calendar. A calendar of working days or shifts that establishes those dates on which schedule activities are worked and nonworking days that determine those dates on which schedule activities are idle. Typically defines holidays, weekends, and shift hours. See also resource calendar.

Project Charter [Output/Input]. A document issued by the project initiator or sponsor that formally authorizes the existence of a project, and provides the project manager with the authority to apply organizational resources to project activities.

Project Communications Management [Knowledge Area]. Project Communications Management includes the processes required to ensure timely and appropriate generation, collection, distribution, storage, retrieval, and ultimate disposition of project information.

Project Cost Management [Knowledge Area]. Project Cost Management includes the processes involved in estimating, budgeting, and controlling costs so that the project can be completed within the approved budget.

Project Human Resource Management [Knowledge Area]. Project Human Resource Management includes the processes that organize and manage the project team.

Project Initiation. Launching a process that can result in the authorization of a new project.

Project Integration Management [Knowledge Area]. Project Integration Management includes the processes and activities needed to identify, define, combine, unify, and coordinate the various processes and project management activities within the Project Management Process Groups.

Project Life Cycle. A collection of generally sequential project phases whose name and number are determined by the control needs of the organization or organizations involved in the project. A life cycle can be documented with a methodology.

Project Management. The application of knowledge, skills, tools, and techniques to project activities to meet the project requirements.

Project Management Body of Knowledge. An inclusive term that describes the sum of knowledge within the profession of project management. As with other professions, such as law, medicine, and accounting, the body of knowledge rests with the practitioners and academics that apply and advance it. The complete project management body of knowledge includes proven traditional practices that are widely applied and innovative practices that are emerging in the profession. The body of knowledge includes both published and unpublished materials. This body of knowledge is constantly evolving. PMI's PMBOK ® Guide identifies that subset of the project management body of knowledge that is generally recognized as good practice.

Project Management Information System (PMIS) [Tool]. An information system consisting of the tools and techniques used to gather, integrate, and disseminate the outputs of project management processes. It is used to support all aspects of the project from initiating through closing, and can include both manual and automated systems.

Project Management Knowledge Area. An identified area of project management defined by its knowledge requirements and described in terms of its component processes, practices, inputs, outputs, tools, and techniques.

Project Management Office (PMO). An organizational body or entity assigned various responsibilities related to the centralized and coordinated management of those projects under its domain. The responsibilities of a PMO can range from providing project management support functions to actually being responsible for the direct management of a project.

Project Management Plan [Output/Input]. A formal, approved document that defines how the project is executed, monitored, and controlled. It may be a summary or detailed and may be composed of one or more subsidiary management plans and other planning documents.

Project Management Process Group. A logical grouping of project management inputs, tools and techniques, and outputs. The Project Management Process Groups include initiating processes, planning processes, executing processes, monitoring and controlling processes, and closing processes. Project Management Process Groups are not project phases.

Project Management System [Tool]. The aggregation of the processes, tools, techniques, methodologies, resources, and procedures to manage a project.

Project Management Team. The members of the project team who are directly involved in project management activities. On some smaller projects, the project management team may include virtually all of the project team members.

Project Manager (PM). The person assigned by the performing organization to achieve the project.

Project Organization Chart [Output/Input]. A document that graphically depicts the project team members and their interrelationships for a specific project.

Project Phase. A collection of logically related project activities, usually culminating in the completion of a major deliverable. Project phases are mainly completed sequentially, but can overlap in some project situations. A project phase is a component of a project life cycle. A project phase is not a Project Management Process Group.

Project Procurement Management [Knowledge Area]. Project Procurement Management includes the processes to purchase or acquire the products, services, or results needed from outside the project team to perform the work.

Project Quality Management [Knowledge Area]. Project Quality Management includes the processes and activities of the performing organization that determine quality policies, objectives, and responsibilities so that the project will satisfy the needs for which it was undertaken.

Project Risk Management [Knowledge Area]. Project Risk Management includes the processes concerned with conducting risk management planning, identification, analysis, responses, and monitoring and control on a project.

Project Schedule [Output/Input]. The planned dates for performing schedule activities and the planned dates for meeting schedule milestones.

Project Schedule Network Diagram [Output/Input]. Any schematic display of the logical relationships among the project schedule activities. Always drawn from left to right to reflect project work chronology.

Project Scope. The work that must be performed to deliver a product, service, or result with the specified features and functions.

Project Scope Management [Knowledge Area]. Project Scope Management includes the processes required to ensure that the project includes all the work required, and only the work required, to complete the project successfully.

Project Scope Statement [Output/Input]. The narrative description of the project scope, including major deliverables, project assumptions, project constraints, and a description of work, that provides a documented basis for making future project decisions and for confirming or developing a common understanding of project scope among the stakeholders.

Project Time Management [Knowledge Area]. Project Time Management includes the processes required to manage the timely completion of a project.

Projectized Organization. Any organizational structure in which the project manager has full authority to assign priorities, apply resources, and direct the work of persons assigned to the project.

Quality. The degree to which a set of inherent characteristics fulfills requirements.

Quality Management Plan [Output/Input]. The quality management plan describes how the project management team will implement the performing organization's quality policy. The quality management plan is a component or a subsidiary plan of the project management plan.

Regulation. Requirements imposed by a process, or service characteristics, including compliance.

Reserve. A provision in the project management plan to mitigate cost and/or schedule risk. Often used with a modifier (e.g., management reserve, contingency reserve) to provide further detail on what types of risk are meant to be mitigated.

Reserve Analysis [Technique]. An analytical technique to determine the essential features and relationships of components in the project management plan to establish a reserve for the schedule duration, budget, estimated cost, or funds for a project.

Resource. Skilled human resources (specific disciplines either individually or in crews or teams), equipment, services, supplies, commodities, material, budgets, or funds.

Resource Breakdown Structure. A hierarchical structure of resources by resource category and resource type used in resource leveling schedules and to develop resource-limited schedules, and which may be used to identify and analyze project human resource assignments.

Resource Leveling [Technique]. Any form of schedule network analysis in which scheduling decisions (start and finish dates) are driven by resource constraints (e.g., limited resource availability or difficult-to-manage changes in resource availability levels).

Responsibility Assignment Matrix (RAM) [Tool]. A structure that relates the project organizational breakdown structure to the work breakdown structure to help ensure that each component of the project's scope of work is assigned to a person or team.

Result. An output from performing project management processes and activities. Results include outcomes (e.g., integrated systems, revised process, restructured organization, tests, trained personnel, etc.) and documents (e.g., policies, plans, studies, procedures, specifications, reports, etc.). Contrast with product. See also deliverable.

Risk. An uncertain event or condition that, if it occurs, has a positive or negative effect on a project's objectives.

Risk Acceptance [Technique]. A risk response planning technique that indicates that the project team has decided not to change the project management plan to deal with a risk, or is unable to identify any other suitable response strategy.

Risk Avoidance [Technique]. A risk response planning technique for a threat that creates changes to the project management plan that are meant to either eliminate the risk or to protect the project objectives from its impact.

Risk Breakdown Structure (RBS) [Tool]. A hierarchically organized depiction of the identified project risks arranged by risk category and subcategory that identifies the various areas and causes of potential risks. The risk breakdown structure is often tailored to specific project types.

Risk Category. A group of potential causes of risk. Risk causes may be grouped into categories such as technical, external, organizational, environmental, or project management. A category may include subcategories such as technical maturity, weather, or aggressive estimating.

Risk Management Plan [Output/Input]. The document describing how project risk management will be structured and performed on the project. It is contained in or is a subsidiary plan of the project management plan. Information in the risk management plan varies by application area and project size. The risk management plan is different from the risk register that contains the list of project risks, the results of risk analysis, and the risk responses.

Risk Mitigation [Technique]. A risk response planning technique associated with threats that seeks to reduce the probability of occurrence or impact of a risk to below an acceptable threshold.

Risk Register [Output/Input]. The document containing the results of the qualitative risk analysis, quantitative risk analysis, and risk response planning. The risk register details all identified risks, including description, category, cause, probability of occurring, impact(s) on objectives, proposed responses, owners, and current status.

Risk Tolerance. The degree, amount, or volume of risk that an organization or individual will withstand.

Risk Transference [Technique]. A risk response planning technique that shifts the impact of a threat to a third party, together with ownership of the response.

Schedule. See project schedule and see also schedule model.

Schedule Baseline. A specific version of the schedule model used to compare actual results to the plan to determine if preventive or corrective action is needed to meet the project objectives.

Schedule Compression [Technique]. Shortening the project schedule duration without reducing the project scope. See also crashing and fast tracking.

Schedule Management Plan [Output/Input]. The document that establishes criteria and the activities for developing and controlling the project schedule. It is contained in, or is a subsidiary plan of, the project management plan.

Schedule Network Analysis [Technique]. The technique of identifying early and late start dates, as well as early and late finish dates, for the uncompleted portions of project schedule activities. See also critical path method, critical chain method, and resource leveling.

Schedule Performance Index (SPI). A measure of schedule efficiency on a project. It is the ratio of earned value (EV) to planned value (PV). The SPI = EV divided by PV.

Schedule Variance (SV). A measure of schedule performance on a project. It is the difference between the earned value (EV) and the planned value (PV). SV = EV minus PV.

Scope. The sum of the products, services, and results to be provided as a project. See also project scope and product scope.

Scope Baseline. An approved specific version of the detailed scope statement, work breakdown structure (WBS), and its associated WBS dictionary.

Scope Change. Any change to the project scope. A scope change almost always requires an adjustment to the project cost or schedule.

Scope Creep. Adding features and functionality (project scope) without addressing the effects on time, costs, and resources, or without customer approval.

Scope Management Plan [Output/Input]. The document that describes how the project scope will be defined, developed, and verified and how the work breakdown structure will be created and defined, and that provides guidance on how the project scope will be managed and controlled by the project management team. It is contained in or is a subsidiary plan of the project management plan.

Sensitivity Analysis. A quantitative risk analysis and modeling technique used to help determine which risks have the most potential impact on the project. It examines the extent to which the uncertainty of each project element affects the objective being examined when all other uncertain elements are held at their baseline values. The typical display of results is in the form of a tornado diagram.

Sequence Activities [Process]. The process of identifying and documenting relationships among the project activities.

Slack. Also called fl oat. See total fl oat and free fl oat.

Specification. A document that specifies, in a complete, precise, verifiable manner, the requirements, design, behavior, or other characteristics of a system, component, product, result, or service and, often, the procedures for determining whether these provisions have been satisfied. Examples are: requirement specification, design specification, product specification, and test specification.

Sponsor. The person or group that provides the financial resources, in cash or in kind, for the project.

Stakeholder. Person or organization (e.g., customer, sponsor, performing organization, or the public) that is actively involved in the project, or whose interests may be positively or negatively affected by execution or completion of the project. A stakeholder may also exert influence over the project and its deliverables.

Start Date. A point in time associated with a schedule activity's start, usually qualified by one of the

Statement of Work (SOW). A narrative description of products, services, or results to be supplied.

Strengths, Weaknesses, Opportunities, and Threats (SWOT) Analysis. This information gathering technique examines the project from the perspective of each project's strengths, weaknesses, opportunities, and threats to increase the breadth of the risks considered by risk management.

Subphase. A subdivision of a phase.

Subproject. A smaller portion of the overall project created when a project is subdivided into more manageable components or pieces.

Successor Activity. The schedule activity that follows a predecessor activity, as determined by their logical relationship.

Summary Activity. A group of related schedule activities aggregated at some summary level, and displayed/ reported as a single activity at that summary level. See also subproject and subnetwork.

Team Members. See project team members.

Technique. A defined systematic procedure employed by a human resource to perform an activity to produce a product or result or deliver a service, and that may employ one or more tools.

Template. A partially complete document in a predefined format that provides a defined structure for collecting, organizing, and presenting information and data.

Three-Point Estimate [Technique]. An analytical technique that uses three cost or duration estimates to represent the optimistic, most likely, and pessimistic scenarios. This technique is applied to improve the accuracy of the estimates of cost or duration when the underlying activity or cost component is uncertain.

Threshold. A cost, time, quality, technical, or resource value used as a parameter, and which may be included in product specifications. Crossing the threshold should trigger some action, such as generating an exception report.

Time and Material (T&M) Contract. A type of contract that is a hybrid contractual arrangement containing aspects of both cost-reimbursable and fixed-price contracts. Time and material contracts resemble cost reimbursable type arrangements in that they have no definitive end, because the full value of the arrangement is not defined at the time of the award. Thus, time and material contracts can grow in contract value as if they were cost-reimbursable-type arrangements. Conversely, time and material arrangements can also resemble fixed-price arrangements. For example, the unit rates are preset by the buyer and seller, when both parties agree on the rates for the category of senior engineers.

Tool. Something tangible, such as a template or software program, used in performing an activity to produce a product or result.

Total Float. The total amount of time that a schedule activity may be delayed from its early start date without delaying the project finish date, or violating a schedule constraint. Calculated using the critical path method technique and determining the difference between the early finish dates and late finish dates. See also free float.

Trend Analysis [Technique]. An analytical technique that uses mathematical models to forecast future outcomes based on historical results. It is a method of determining the variance from a baseline of a budget, cost, schedule, or scope parameter by using prior progress reporting periods' data and projecting how much that parameter's variance from baseline might be at some future point in the project if no changes are made in executing the project.

Triggers. Indications that a risk has occurred or is about to occur. Triggers may be discovered in the risk identification process and watched in the risk monitoring and control process. Triggers are sometimes called risk symptoms or warning signs.

Variance Analysis [Technique]. A method for resolving the total variance in the set of scope, cost, and schedule variables into specific component variances that are associated with defined factors affecting the scope, cost, and schedule variables.

Verify Scope [Process]. The process of formalizing acceptance of the completed project deliverables.

Virtual Team. A group of persons with a shared objective who fulfill their roles with little or no time spent meeting face to face. Various forms of technology are often used to facilitate communication among team members. Virtual teams can be comprised of persons separated by great distances.

Voice of the Customer. A planning technique used to provide products, services, and results that truly reflect customer requirements by translating those customer requirements into the appropriate technical requirements for each phase of project product development.

Work Authorization. A permission and direction, typically written, to begin work on a specific schedule activity or work package or control account. It is a method for sanctioning project work to ensure that the work is done by the identified organization, at the right time, and in the proper sequence.

Work Breakdown Structure (WBS) [Output/Input]. A deliverable-oriented hierarchical decomposition of the work to be executed by the project team to accomplish the project objectives and create the required deliverables. It organizes and defines the total scope of the project.

Work Package. A deliverable or project work component at the lowest level of each branch of the work breakdown structure. See also control account.

Project Management Foundations continues . . .

Excerpts from BOOK 2
Division 4

Project Management Stonewalls

Chapter 9

Integration Management

- Develop Project Charter
- Develop Preliminary Scope Statement
- Develop Project Management Plan
- Direct Execution of the Project
- Monitor and Control Project
- Integrated Change Control
- Close Project

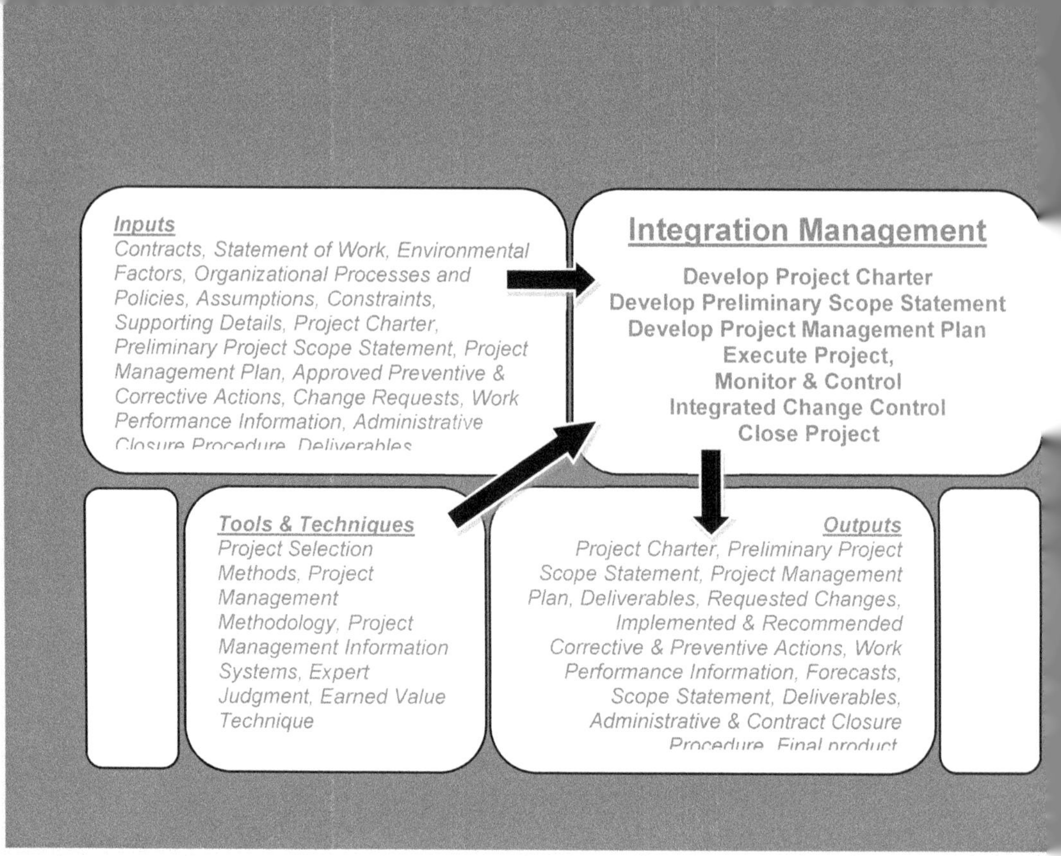

Inputs
Contracts, Statement of Work, Environmental Factors, Organizational Processes and Policies, Assumptions, Constraints, Supporting Details, Project Charter, Preliminary Project Scope Statement, Project Management Plan, Approved Preventive & Corrective Actions, Change Requests, Work Performance Information, Administrative Closure Procedure, Deliverables

Integration Management

Develop Project Charter
Develop Preliminary Scope Statement
Develop Project Management Plan
Execute Project,
Monitor & Control
Integrated Change Control
Close Project

Tools & Techniques
Project Selection Methods, Project Management Methodology, Project Management Information Systems, Expert Judgment, Earned Value Technique

Outputs
Project Charter, Preliminary Project Scope Statement, Project Management Plan, Deliverables, Requested Changes, Implemented & Recommended Corrective & Preventive Actions, Work Performance Information, Forecasts, Scope Statement, Deliverables, Administrative & Contract Closure Procedure, Final product

Figure 9.1 – Inputs, Tools, & Outputs of Project Integration Management

The concept of project integration can be likened to systems integration, which involves the amalgamation of different part's phases and processes of a project into one functioning project system. Effectively identifying, defining, linking, and coordinating all aspects of a project make for easy management. You, as the project integrator, will utilize a variety of tools, techniques, and skills (Figure 9.1) to aggregate parts and sub-parts, processes and activities of a project to deliver the project goal. The function of project integration management cuts across all project process groups from initiation to closing. These processes are discrete components with interfaces that interact and overlap.

Project integration is fundamental because these individual processes that interact in ways that cannot be completely delineated, need to be interconnected or interlocked. Any manifestation of inadequacy in the ability to hold these processes together can grant each or any process an uncontrollable life of its own. You do not want this; no project manager does. Therefore, it is important that you possess the capability required to identify, define, combine, unify, and coordinate various activities and processes of project management. To this end, different project managers employ different project coordination methods that apply knowledge, skills, and processes in varying degrees and order to accomplish project goals.

Consider the high intelligence, keen eyesight, sense of touch, slow and fast locomotion, defense strategies, and more importantly the coordinative mechanisms of an eight-arm octopus. It is claimed that an octopus has a highly complex nervous system, which a third of its neurons—information processing and transmitting cells—is located in its brain. The other two-thirds are located in the nerve cords of its arms, which demonstrate a range of complex reflex actions.

Similar to the brain of an octopus, project integration management is not the only important information processing knowledge area of project management. However, it is central to coordinating the 14 interacting arms—nine knowledge areas and five process groups—of project management. The octopus chooses which and the number of arms to walk on, how fast to swim, how slow to crawl, or what defense mechanism to adopt in specific situations. As a project integration manager, you make choices at any point in time about how to deploy and where to concentrate resources and effort in anticipation of issues before they become critical. This also means that you have to be aware of the various other activities performed in the process of completing a project.

Project Management Area & Process	Project Management Process Groups & Functions				
	Initiating	Planning	Executing	Monitoring & Controlling	Closing
Integration	Develop Project Charter * Develop Preliminary Scope Statement	Develop Project Management Plan	Direct Execution of Project	Monitor & Control Project * Integrated Change Control	Close Project

Figure 9.2 – Project Integration Management Process and Process Groups Interaction

In critical circumstances with predators, the octopus manifests an ability to release an inky fluid to confuse the enemy, changes it's coloration to remain indiscernible from a surrounding environment, or sever one of its limbs to escape from a predators grip. Your unification, consolidation, articulation, and integrative efforts in project integration will require making trade-offs among competing objectives and alternatives.

These efforts are vital to successfully accomplish your project goals and ...

Milton Keynes UK
Ingram Content Group UK Ltd.
UKHW011516230124
436534UK00001B/147